PRINCE OF PERIL

Otis Adelbert Kline

Prince of Peril

Table of Contents

Foreword
Chapter I
Chapter II
Chapter III
Chapter IV
Chapter V
Chapter VI
Chapter VII
Chapter VIII
Chapter IX
Chapter X
Chapter XI
Chapter XII
Chapter XIII
Chapter XIV
Chapter XV
Chapter XVI
Afterword

Foreword

Many people have asked me how I came to write "The Swordsman of Mars," "The Outlaws of Mars," and "The Planet of Peril," and have wondered why the character of Dr. Morgan appears in all of them. "It was all right for the first story," one reader complained, "but it begins to get a bit thick the third time. I hope you're not going to do it again." Another thought that Dr. Morgan really belonged in the series, but that there wasn't enough of him; I should justify his continuance by having him play a more important role in the plot.

As an author, I agree with both of these critics. Dr. Morgan either should have been dropped, or should have a more active and vital role; and I certainly would have taken one of these alternatives in the second novel, "Outlaws of Mars," were this series really my own to work out as I pleased.

You see, while the name "Dr. Morgan" is fictitious, the character is not. It was quite by accident that I literally dropped in on him one day while deer-hunting in the mountains. It was a cloudy day, and I lost my bearings. I'd been foolish enough to forget my compass, so I climbed the highest prominence to orient myself.

If you have ever met me, you will know that these were not tremendous mountains. Now that I'm letting you in on a long-kept secret, I must confess to further deception. If you will re-read the opening chapters of the preceding books, you will see that while I've given the impression that Dr. Morgan's retreat was amidst high mountains, I've never said anything definite about the height. There were high enough for my own purposes of sport and exercise, and Dr. Morgan's purposes of isolation, but you may have been led to overestimate their eminence.

I had all but reached the summit I was approaching, when my feet suddenly slipped from under me. Gun and all, I crashed through something which felt and sounded like glass, and struck a hard, concrete floor. My right leg crumpled under me, and all went black.

When I regained consciousness I thought I was in a hospital, for two men in white garments were working over me.

The younger man I took to be an interne. The other was indeed a doctor, as I was to learn. He was of gigantic stature, but well–proportioned and athletic, and of most striking appearance. His forehead was far higher than any other I had ever seen, bulging outward so that his shaggy eyebrows, which grew completely together above the bridge of his aquiline nose, half concealed his small, glittering, beady eyes. His close–cropped, sharply pointed beard, in which a few gray hairs were in evidence, proclaimed him as probably past middle age.

When he had finished bandaging my fractured leg, which throbbed unmercifully, he dismissed his assistant, called me by name, and introduced himself. I am not yet free to divulge his true identity, so I shall continue to call him "Dr. Morgan."

"What hospital is this," I asked, "and how did you find me?"

"You are not in a hospital," he replied in his booming bass voice, "but still on the mountain in my retreat. My men are now replacing the skylight through which you fell."

For nearly a month I convalesced in the secret, perfectly–camouflaged observatory. When he learned that I was an author (he had learned my name from the mundane process of looking through my wallet) he asked permission to question me under hypnosis, promising to explain when he had finished, and assuring me that I need not worry about anything he would ask me.

There are some human beings who inspire you with trust almost upon first sight. Dr. Morgan was such a person. I agreed; and I learned later that, had he not been trustworthy, it would have been very easy for him to have tricked me into agreement. Actually, he would not have done it without my full consent, honestly gained.

"I must ask your forgiveness," he said, after the session. "While my impression of you was that you were both honest and reliable, I had to be sure that you did not have particular character weaknesses through which you could be easily led to betray confidences you really meant to keep. I have some material which would be ideal for the sort of stories you write, but it is vital that certain aspects of what you will learn do not become public knowledge. Without these, few readers will suspect that what you will write is anything but very imaginative romance, and those few will not be able to ascertain more without facts which I now am confident you won't reveal."

He stroked his beard. "I could, of course, with your consent, doubly insure security by putting you under hypnotic inhibition—you would not remember what you were not supposed to reveal. But this is a risky process, not one hundred percent certain, and might have undesirable side-effects upon you."

"I'll go along with your judgment on this," I told him.

In the days that followed I learned about Dr. Morgan's studies of parapsychology, particularly in telepathy. I had done some reading in this line myself, so knew something of the general theory—that the communication of thoughts or ideas or moods from one mind to another without the use of any physical medium whatever was not influenced or hampered by either time or space.

Dr. Morgan had worked on telepathy for many years in his spare time, when he was in practice; but on his retirement, he tried a different track. "I had to amend the theory," he explained. "I decided that it would be necessary to build a device which would pick up and amplify thought waves. And even this would have failed had my machine not caught the waves projected by another machine, which another man had built to amplify and project them."

Now I had been a devotee of imaginative fiction for many years, and had often thought of turning my hand to writing it. I prided myself on having a better than usual imagination; yet, I did not think of the implications of the theory of telepathy when Dr. Morgan told me that the man who built the thought-projector was on Mars. While I deferred to no one in my fondness for Edgar Rice Burroughs's stories of John Carter and others on Barsoom, I was well aware of the fact that what we knew of the planet Mars made his wonderful civilization on that planet quite impossible. I said as much, going into facts and figures.

"Of course, we won't really know for sure about the exact conditions there unless we land on Mars. But still we know enough to make Burroughs's

Mars probability zero," I concluded.

Dr. Morgan nodded. "Entirely correct," he said. "There is no such civilization on Mars."

He then explained his own incredulity when his machine picked up the thoughts of a man who identified himself as a human being—one Lal Vak, a Martian scientist and psychologist. But Lal Vak was no less incredulous when Dr. Morgan identified himself as a human being and scientist of Earth. For Lal Vak was certain that there could be no human civilization on Earth, and cited facts and figures to prove it.

And that was the clue. Both Dr. Morgan and Lal Vak were correct. Neither man could possibly exist on the world he claimed to inhabit—if both were living in the same area of space–time. But Lal Vak's description of Earth was a valid description of the third planet from the sun as it existed millions of years ago.

"I have read many weird and fantastic stories," Dr. Morgan said, "as have you. Some of them have given me a most eerie feeling—but nothing to compare with my feelings upon talking with a man who has been dead millions of years, of whose civilization there may now linger not so much as a single trace."

This was the beginning. Dr. Morgan brought me several thick typewritten manuscripts which he had bound separately, and I read therein the stories of Harry Thorne, of Morgan's own nephew, Jerry, and of Robert Grandon. Thus I learned that Lal Vak was the contemporary of a Venusian named Vorn Vangal and that a human civilization had also existed on Venus at this time.

With the aid of Lal Vak, Dr. Morgan had effected transfer of personalities between two Martians and two Earthmen, whose physical and brain–pattern make–up were similar enough to permit such exchange. Through a means which I am still barred from describing in detail, it was possible for Dr. Morgan to keep in rapport with his emissaries on Mars—providing they co–operated. The first man broke contact, and turned out to be a disasterously wrong choice. Thus, Harry Thorne was sent to Mars, to exchange consciousness with a Martian whose body was holding the personality of Frank Boyd, criminal Earthman.

From Vorn Vangal, Dr. Morgan learned the construction and operation of a space–time vehicle, propelled by telekinesis. It was by means of this vehicle that Morgan's nephew, Jerry, went to Mars physically. But something went wrong on the return trip—Dr. Morgan had tried to bring the vehicle back to Earth and his own time, empty, for use to transport an Earthman to Venus later—and the vehicle was lost.

"It might have been possible to build another," Dr. Morgan told me, after I had finished reading about the adventures of his nephew, "but Vorn Vangal and I decided that it would be simpler to use the personality–exchange system, if we could find an Earthman or two who could qualify." He pointed to the other two manuscripts which I was yet to read. "These tell of what happened to the two I sent to Venus: Robert Grandon and Rorgen Takkor."

"Rorgen Takkor—but he's on Mars," I protested. "He's the Zovil of Xancibar…Did something go wrong? A break–up between him and Neva…?"

Dr. Morgan smiled. "No, no, my friend—Harry Thorne is on Mars in the body of Rorgen Takkor. The man who was my assistant for many years, called Harry Thorne, is Rorgen Takkor." He coughed slightly. "Of course, he is now known as Prince Zinlo of Venus."

I smiled. "If we can consider millions of years in the past as 'now.'"

"I am still in contact with him, as with the others who are 'still' alive…At any rate, Rorgen Takkor asked me if he could go to Venus; he was getting tired of Earth, and of course he could not return to Mars. He was fascinated with what Vorn Vangal told me of the Venusian civilization and was sure he'd feel more at home there, however strange it might be. I'd say it would be roughly analogous to the case of a crusader from 12th Century England transported and settled down into a remote part of Islam, where there was not and probably never would be direct contact with his native civilization."

So "Harry Thorne," and an Earthman named Robert Grandon went to Venus.

Here were four distinct stories, and Dr. Morgan went over them with me, indicating what parts of them might be used for novels, and what had best not be related in detail, or omitted entirely.

I have told you the story of Robert Grandon in "The Planet of Peril," and those of you who have read it will recall that Harry Thorne and Grandon met in the closing episodes of the story. You may remember that Grandon asked Thorne to tell him of his adventures between the time of Thorne's arrival on Venus and this meeting, as it was plain that much had happened and that the other man had found his place and the woman of his heart's desire. Before Thorne could tell the story, they were interrupted by announcements that their airship had arrived at Vernia's capital.

Actually, the record shows that Thorne did tell his story to Grandon later, during the visit—although like nothing in the detail present in Dr. Morgan's records. But it was impossible to give even so brief an outline in this place. It had no bearing on the story of Robert Grandon and his rise on Venus, his winning of Vernia, and the defeat and death of the traitor, Prince Destho. I decided to omit it entirely, leaving it for another novel.

So now I offer you the story of Harry Thorne—and, with your permission, I shall stop calling him "Harry Thorne." This is the story of Rorgen Takkor's adventures on Venus, Rorgen Takkor, born on Mars, transferred to Earth for a decade, and finally finding his career and place on Venus.

The Author.

Chapter I

"Good–bye, men and good luck to you."

My awakening, after I lay down on the cot in Dr. Morgan's observatory, was quite sudden and startling. It seemed that not more than a few seconds had elapsed since I had heard the doctor's parting words to Grandon and myself.

I opened my eyes and sat up abruptly with an inexplicable sense of impending danger. My first glimpse of my surroundings convinced me that I had indeed arrived on Venus. The magnificent riot of vegetation surrounding me—vegetation the like of which I had not seen on Mars, the red, barren planet of my birth, nor on Earth, the more recent planet of my adoption—was sufficient evidence.

I was seated on a bank of soft, violet–colored moss which sloped gently to a limpid pool at my feet. The feathery fronds of a giant bush–fern arched above my head, some of them dipping to the surface of the water, where they were snapped at from time to time by playful, grotesque, multi–colored amphibians.

I was dressed in garments of shimmering, scarlet material. There was a broad, golden chain–belt about my waist, with a jeweled clasp in front. Riveted to this belt on the right side was an oblong instrument about two feet in length, with a button near the upper end, a small lever on the side, and a tiny hole in the lower end. I had no idea what it was for; but I recognized the weapon which hung at my left side, as it resembled a scimitar. As I was examining the ruby–studded hilt of this beautiful weapon, a noise at my left attracted my attention.

Cautiously, without turning my head, I glanced from the corners of my eyes across a stretch of shrubbery to where a high wall of black stone surrounded this estate, and hid the country beyond. Just on the other side of the wall a tall fern–tree spread its mighty fronds. It must have been the cracking of one of these that had attracted my attention, for a heavy–set individual with a coarse red beard, cut off square below the chin, had climbed out on it to a point where it would no longer sustain his weight, in an effort to reach the top of the wall.

Someone in the shrubbery quite near me called a whispered warning to him—or such I took it to be, for the language was unknown to me, and I could only judge by the tones. The huge intruder was much more agile than he appeared, for he flung an arm over the top of the wall and drew himself up with catlike quickness. As he struck the wall there was a metallic clank which, I saw as soon as he came into full view, was from an edged weapon at his side, quite like my own but with a less ornate hilt and broader blade.

As soon as the red–bearded man reached the top of the wall, the one who had whispered from the bushes cautiously stood up. He was smaller and more wiry than the first, and his beard, which was iron–gray in color, was trimmed in the same manner.

Red–beard tiptoed stealthily along the top of the wall, glancing toward me from time to time as if fearful that I would hear him or turn toward him. Then he leaned out, caught his fingers in a tall cone–shaped growth, swung his sandaled feet out, and descended.

I wondered if it could be possible that these two prowlers were bent on injury to me, a total stranger on Venus. Then it dawned on me that they could easily be mortal enemies of the prince with whom I had exchanged bodies, and that I—so far as their knowledge went—was that prince.

I therefore drew my cutting weapon from its sheath in order to have it ready, and pretended to examine its beautiful, highly polished blade. For several minutes I neither saw nor heard anything of the two prowlers. Then I suddenly glimpsed, reflected on the polished surface of my blade, the red–bearded man standing directly behind me with his weapon upraised for a downward cut that would have sheared my skull from crown to chin. As swords of all kinds had been my principal playthings on Mars, and fencing my favorite amusement on Earth, I did the thing which any swordsman would have done instinctively in the circumstances. I raised the blade of my weapon above my head with a downward slant from hilt to point, and the descending blade of my would–be assassin, deflected by my own, buried itself in the mossy turf on my left.

Springing to my feet, I whirled and attacked.

My opponent proved to be a hammer–and–tongs fighter, no match for superior swordsmanship. I could have killed him any one of a dozen times before he realized that I was playing with him. Then he bawled out lustily, and the wiry fellow with the gray beard came rushing out of the bushes. Not knowing the caliber of the second assailant, I stopped the squawking of the first with a quick neck–cut that laid him low.

The wiry graybeard was much quicker and far more elusive than his huge companion, and I did not play with him. He soon left me the opening I sought, and I stretched him beside his fellow with a bone–shearing cut.

Having ascertained beyond doubt that both of my would–be assassins were dead, I carefully cleaned my blade, sheathed it, and set out to explore my surroundings.

I had been walking for perhaps ten minutes along the mossy bank, when a monster, more hideous than anything I had ever seen or even dreamed existed, emerged from the water and came toward me.

I whipped out my blade as it waddled forward on its thick, bowed legs. Its long, scaly tail dragged in the moss, and its enormous jaws were distended in a grin that disclosed huge, ivory–white tusks. It was so fearsome a thing that, although I am no coward, I knew not whether to stand and fight or take to my heels.

A gust of laughter at my right caused me to turn. I beheld a tall man, apparently of middle age, smiling broadly at me. His garments were of purple, and he wore a beard that had once been black, now slightly streaked with gray, cut off square below the chin. His weapons were similar to mine, though his belt was of silver.

"The 'ikthos' will not harm you," he said in English. "It is one of the garden pets, and hostile only to strangers."

The thing he called an ikthos sniffed at my garments, rubbed its ugly muzzle against my thigh, yawned, and crouched at my feet.

"You are surprised at my knowledge of English," continued my new acquaintance. "After I tell you who you are and were, and also who I am, the thing will not seem so mysterious. You are he who was Rorgen Takkor on Mars, and later Harry Thorne on Earth. You have now become Zinlo, the Torrogi or Imperial Crown Prince of Olba. I am Vorn Vangal, the Olban psychologist, and have been communicating telepathically with Dr. Morgan of Earth for several years."

"I have heard the doctor speak of you often," I replied. "It is a pleasure to meet you, Vorn Vangal."

He acknowledged with a courtly bow. "I have but a few hours to spend with you. Grandon has already arrived on the other side of the planet and will shortly awaken to find himself a princely slave in the marble quarries of Uxpo. I must fly to his assistance. Come with me and see what preparations I have made for you."

I followed Vorn Vangal through the garden. There was a profusion of ornamental trees, shrubs, fungi and jointed grasses, but no flowers or fruits. Patches of gloriously colored water plants of divers odd shapes flourished in the lagoons, and fungi of a thousand types and sizes grew in the moister places.

Though it was without flowers, the garden did not lack color. All the hues of the rainbow were represented in its rankly growing, primitive vegetation. Toadstools as tall as trees bordered several of the lagoons, some of them lemon–yellow, others orange, scarlet, black or brown, and still others of pale, chalky whiteness.

Beautiful statues and statuettes stood here and there, some placed conspicuously, but more of them showing unexpectedly in niches and vine–covered bowers as we moved along.

The garden teemed with bird and animal life. The trees were alive with gay–plumed songbirds that filled the air with their melodious, flute–like notes. Waterfowl, both swimmers and waders, dotted the lagoons, and their cries, though not musical, were far from unpleasant. Amphibians of many species disported themselves in the water or dozed lazily on the banks. I was astonished at sight of a huge yellow frog which must easily have measured more than six feet from nose to toes, blinking contentedly and fearlessly down at me from his seat on an enormous scarlet toadstool.

With our hideous ikthos trailing closely behind us, and from time to time affectionately nosing either Vorn Vangal or me with its cold, moist snout, we presently came before a tall building. It was of black marble, and was my first glimpse of Olban architecture.

Its shape astonished me. I do not believe there was a straight line in the entire structure. Everything was curved. The building stood on a circular foundation, and its walls, instead of mounting skyward in a straight line, bellied outward and then curved in again at the top. The lower structure was surmounted by a second segment, smaller, but of similar shape. This, in turn, supported others, still smaller, up to the top segment, some thirty feet in diameter and no less than six hundred feet from the ground.

We mounted a flight of steps, walked between two uniformed guards who saluted stiffly, and entered a large circular door, where a slave took charge of the ikthos and led him away. After following a broad hallway for some distance we came to a huge pillar. It was in the center of the building, and was decorated on one side with a large oval plate of burnished silver on which was embossed what appeared to be a coat–of–arms. As we stepped before it the plate slid back, revealing a small room within.

At Vangal's invitation I stepped into the small room inside the huge central pillar of the tower, and he followed. As soon as he stood beside me the silver plate slid back across the entrance, a concealed light flashed on somewhere above our heads, and the floor moved upward.

We were in an elevator, of course, but what had started the thing and how was my companion going to stop it when we reached our destination? There were no levers or buttons of any sort. The thing seemed almost human in its movements. Perhaps there was a hidden operator. I voiced my question to Vorn Vangal.

"It is moved by a mechanism which amplifies the power of telekinesis," he said.

I had often heard Dr. Morgan use the word "telekinesis," and knew that it described that mysterious power of the mind which enables psychics to tip tables and lift imponderable objects without physical means. In short, it referred to the direct power of mind over matter.

"I have heard of small objects being moved or lifted by telekinesis," I marveled, "but to lift an elevator! Why, this is amazing!"

"We lift far heavier things than this little car," said Vangal, smiling slightly. "Huge cranes and derricks are operated in the same way. Airships of all sizes from small one–man flyers to huge battleships are moved by it—propelled through the air at speeds ranging from two hundred to one thousand miles an hour."

"But how is that possible?"

"It was made possible by that wonderful invention, the mechanism that amplifies the mind's power. The manufacture of this mechanism is the exclusive secret of the Olban government, and constitutes our defense against aggression from the warlike torro–gats—or empires—surrounding us. If those governments knew the secret, they would build aircraft and use them for conquest. The Olbans, however, are committed to a policy of 'live and let live': We use our wonderful power only for commercial purposes and as a defense against aggression."

We stopped before a metal plate which slid back noiselessly. I stepped out of the car and Vorn Vangal came after me, whereupon the plate slid back in place.

We were in a small, circular hallway around whose walls were metal doors at intervals of about twenty feet. Vangal led the way to one of these doors, pressed a button, and when it slid open, bowed me into a luxuriously furnished suite lighted by enormous circular windows that reached nearly from floor to ceiling.

"This is to be your retreat until my return from Uxpo," he said. "I have been preparing for your coming these many months."

He walked to a beautifully carved table of red wood, and took a thick scroll from a pile neatly stacked on its polished top.

"These are your lessons in patoa, the universal language of Venus. Our patoan name for Venus is Zarovia. Some twenty thousand patoan words are listed here with their pronunciations and English translations. If you will study them carefully until my return it will perhaps be safe for you to leave the Black Tower. Then I can take you to the Red Tower, the Imperial Palace of Olba."

"Am I to infer that it would be unsafe for me to leave the tower at present?"

"The tower and grounds are well guarded," Vorn Vangal replied; "but do not under any circumstances wander beyond the walls. In the course of your walks in the garden, always keep the ikthos with you. He will warn you of lurking assassins, and will fight in your defense."

"He certainly wasn't on the job a short time ago," I said.

"What do you mean?"

I told him of the two assassins.

"The beast must have been lured away by his keeper!" cried Vangal, when I had finished. "The traitor will be dealt with in due time. And those two ruffians—they would be in the pay of Taliboz, of course."

"Who is Taliboz?"

"A man whom I suspect, but against whom I can prove nothing. Nothing! You see—in the course of preparation for your coming, I cast about for an excuse to bring your predecessor here in order that His Imperial Majesty, Emperor Hadjez, might not learn that his son Zinlo was changing places with an Earthman. A ready-made excuse presented itself when word came through the intelligence department of the government that there was a plot on foot to assassinate the male members of the imperial family.

"I immediately suggested that Prince Zinlo be brought here until the plotters could be taken and executed. His majesty readily consented, thus making it possible for me to obtain a quiet retreat for you in which you could learn something of the language and customs of Olba, and at the same time be guarded from danger.

"The plotters have not been apprehended, but I am firmly of the opinion that Taliboz is the ringleader. They have already made an attempt on the life of the Emperor and escaped with the loss of a single man. You can see how you would be exposed by going out unguarded."

"I'm willing to stay here for a while," I replied, "for there is no question about my having to learn this language, patoa, sooner or later. But once I learn your language you won't catch me staying behind walls on account of a few assassins. I was born on Mars, where men do not stay indoors to avoid their enemies; and though I am not familiar with your weapons, I believe I can give some account of myself with this cutting implement at my side if attacked."

"No doubt you can," replied Vangal, "although the two ruffians you killed were probably clumsy fighters. But please bear in mind that you are the Torrogi of Olba—the crown prince—and that your life is not yours to throw away heedlessly."

"Don't ever think I'm going to throw it away," I said. "The man who gets it will have to put up a scrap."

"You might be shot from ambush with a tork."

"A tork?"

"You are wearing one attached to your belt."

Vangal explained the use of the oblong instrument at my side. It was about two feet long and shaped like a carpenter's level. A rivet passed completely through it, about eight inches from the top, fastening it to the belt in such a way that it could be tilted at any angle or pointed in any direction by moving the body.

He pressed a small lever on the side and removed two clips, explaining that one was a gas clip containing a thousand rounds of condensed explosive gas, while the other was a bullet clip which held a thousand rounds of needle–like glass projectiles. These projectiles, he said, were filled with a poison that would paralyze man or beast almost instantly, though the paralysis was only temporary. Other projectiles, he explained, were filled with deadly poison, and still others with explosives. The effective range, he stated, was equal to about ten Earth–miles.

He led me to a window which was open.

"I have prepared a target for you," he said. "You will need to practice with the tork if you are to be able to defend yourself on this planet. Do you see that large white plate against the wall at the other end of the garden?"

Yes."

"I had it erected for your use. It is coated with a substance that will combine with the poison in your tork bullets, emitting a green gas. If you see a green spot appear momentarily on the target you will know that you have registered a hit."

I was eager to try this new weapon, and Vangal, smiling at my eagerness, loaded it for me and showed me how to hold it when pressing the button which fired the gas in the chamber by means of an electric spark. It fed new bullets automatically, he explained.

I confidently fired at the target and waited for a green spot to appear. It remained white. Again I fired with the same result.

"You will need considerable practice," said Vangal. "I am not accounted much of a marksman, but watch."

He fired his tork and a green spot appeared in the center of the target. Then, with no apparent effort, he planted a ring of green spots around it.

When the spots had disappeared I tried again, and managed to hit the target once out of five shots.

"Now let me see what you can do with the scarbo," Vangal said.

"The what?"

"That cutting instrument at your side."

"Oh ho, friend Vangal!" I thought. "You won't find me utterly helpless with this weapon."

He drew his scarbo and I mine. Thinking to best me as easily as he had with the tork, he made as if he would lay my head open.

I parried the blow with ease, then whirled his blade on mine with a movement so sudden that, strong as he was, it flew from his grasp and flashing over his head, clanked in the corner behind him.

"Body of Thorth!" he exclaimed. "That is a marvelous trick!"

I recovered his weapon and handed it to him laughingly.

"On Mars I was raised on a diet of swords," I replied.

"Then I suggest that you confine your efforts to target practice and a mastery of patoa," said Vanga. "I must leave you now to go to the assistance of Grandon. My flyer is on the roof. Would you care to see me off?"

"Assuredly."

I followed him into the elevator.

Chapter II

THE ELEVATOR stopped at the floor of the top segment, and we mounted thence to the roof by a spiral stairway. Two guards armed with torks, scarbos and broad–bladed spears, saluted when we appeared. The roof was made of the same material as the walls, and the slabs of black marble were fitted together so cunningly that the joints were all but concealed. It was circled by a four foot wall perforated on the floor level at intervals to carry off the heavy Zarovian rains.

There were four Olban airships on the roof. I examined the nearest one with interest. It was shaped like a small metal duck–boat about ten feet in length and three in the beam. The cockpit was covered with a glass dome in the back of which was a small door. Within this dome I could see an assortment of levers, buttons and knobs, and the cushioned seat for the driver. The thing that amazed me the most was the fact that it was not equipped with planes, rudder or propeller.

Vangal turned to me. "You seem astonished at our airships."

"They certainly do not resemble any aircraft I have previously seen."

"We have no need of planes, propellers or rudders for this type of flyer," he went on. "As I told you, it is raised, lowered, turned, or moved in any desired direction by amplified mindpower. The amplifying mechanism is under the round bump on the forward deck. The small lids that you see fore and aft conceal safety parachutes. That rectangular protuberance from the front of the cab is a mattork, a weapon operated on the same principle as a tork, but with a greater range and firing much heavier projectiles."

"You told me that the Olban government alone possessed the secret for manufacturing these flying mechanisms," I said. "Suppose one should be forced to land in hostile territory. The craft would then, in all probability, fall into the hands of your enemies, and they could thus easily take the mechanism apart and duplicate it."

"That danger has been foreseen. A vial of powerful acid has been placed in the mechanism of each Olban craft in such a way that it will be immediately broken if tampered with. The acid thus released in the secret mechanism will instantly destroy it."

"Certainly a far-sighted provision," I remarked.

"It has kept us at peace with our neighbors for many centuries," replied Vangal. "I dislike leaving you thus precipitately, but the time has come for departure."

So saying, he opened the door in the back of the cab and entered. After a hurried examination of the control levers and the cannon-like mattork, he said: "Farewell. Study diligently, practice assiduously, and be ever on your guard against assassins."

"If I catch any prowling about I'll practice on them instead of the target. Farewell, and a safe and pleasant journey to you."

The little craft rose slowly at first, then, gradually gathering momentum, it shot to a height of a half mile or more, sped away with amazing rapidity, and was soon lost to view.

I walked to the edge of the wall and looked over. The roof was at least six hundred feet from the ground, though the drop from battlement to battlement was only about sixty feet. Far to the northward I descried a city of circular buildings, in the center of which towered an immense red structure similar in design to the one on which I stood, but at least twice as tall.

This must be the Red Tower of which Vorn Vangal had spoken—the Imperial Palace of Olba. The city walls formed a circle, broken at each point of the compass by a tower which evidently covered a gate.

The countryside, as far as I could see, was divided into well-kept farms on each of which was a round building, probably the home of the owner. People were working in the fields, and here and there I saw men driving huge, grotesque beasts hitched to plows or cultivators.

The animals, which I afterward learned were called thirpeds, were great hairless pachyderms; they stood about eight feet at the shoulder, and weighed four to five tons apiece when full grown. They had huge heads and mouths, sharp–pointed long ears, and relatively thin necks almost half as long as their bodies. They moved with a lumbering gait that reminded me of elephants.

The plants under cultivation were fungi of various kinds, and several varieties of bush–ferns.

A smoothly paved road, straight as an arrow, led from the south gate of Olba past the tower on which I stood, and thence to the great, crescent–shaped Olban harbor of Tureno. This was the marine gateway of the capital, whence Emperor Hadjez sent his mighty fleet of trading vessels out over the rolling, steel–blue waters of the mighty Ropok Ocean.

Along this straight, smooth road rumbled great, one–wheeled carts drawn by thirpeds. The body of a Zarovian cart is inside the huge single wheel that carries it, being suspended on an inner idling wheel that keeps it from turning when the outer wheel revolves. There were also one–wheeled motor–driven vehicles that moved over the road with great speed. I saw some with wheels more than twenty feet in diameter, making all of a hundred Earth miles an hour.

One of the guards accompanied me down the telekinetic elevator, which I had not learned to operate, conducted me to the suite Vangal had prepared for me, and bowing low, with right hand extended palm downward, left me alone. I could hear him pacing back and forth in the hall while I studied the patoa scrolls.

As I pored over the translations and pronunciations with keen interest, it seemed to me that I was reading something I had known well, but had forgotten. I tested myself on this and found, to my surprise, that having once read and pronounced a patoan word, I had learned it.

When I told Vorn Vangal about it afterward, he explained that this was because the brain of Zinlo, which had become mine, knew all of these things already. The subjective mind, having once received an impression, records it forever. Thus, having only to tap my subjective mind, I learned instantly. It amazed and overjoyed me.

Long before the afternoon had waned, I had mastered the entire group of lessons which Vorn Vangal had prepared for me. I was eagerly reading a Zarovian book on natural history, when the advent of sudden darkness, so common in tropical and semi–tropical Venus, interrupted my studies. A rap sounded at the door.

"Enter," I said in patoa, eager to try my newly mastered language.

The door slid open, framing the figure of my guard in silhouette against the lighted hall. He entered and pressed a button, flooding the room with soft light. I could not see the points from which the radiance emanated, so cleverly were the fixtures concealed.

"Your Highness's dinner," announced the guard.

Two slaves entered, bearing a huge double–decked tray laden with at least fifty different dishes. A third followed with a small table, and a fourth with gold service and scarlet napery.

Fish, flesh, and fowl were set before me, as well as numerous dishes concocted from mushrooms and other fungi, and countless others whose origin I could not fathom. There was also a colorless, pleasant–tasting beverage which I afterward learned was called "kova," served hot in small bowls. I found it fully as stimulating as strong wine, though with a slightly different effect.

Having dined as became a prince of Olba, I turned once more to my studies.

Late in the evening a second knock sounded at my door, and a new guard admitted a man who was evidently my valet. He busied himself in the adjoining room for a few minutes, then entered and, bowing before me, announced that my bedchamber was ready.

I entered, to behold a sleeping shelf that curved out from the wall like the nest of a cave–swallow. A scarlet canopy fringed with gold projected above it, and the downy, silken coverlets—scarlet lined with golden yellow—had been turned back invitingly.

My valet brought my scarlet sleeping garments, and I wondered at the preponderance of this color; later, I learned that throughout Zarovia scarlet is the exclusive color of royalty.

Though I had grown drowsy over my studies, the novelty of my situation kept me awake. After several hours, I managed to drift off, only to be awakened by a sharp, metallic clang.

The sound seemed to come from the direction of the battlement outside my window, and I listened breathlessly for a repetition. As it was not repeated, I decided that it could have no alarming significance, and was once more composing myself for slumber when I heard a slight rustle as of silken garments only a few feet distant from my head.

Without moving, I opened my eyes and endeavored to penetrate the pitch darkness that enveloped me. Venus has no moon, and in consequence it was fully as dark outside as anywhere in the room; I could not see the window, nor could I have seen any one entering it.

It was plainly evident that there was someone in the room. I thought of Vorn Vangal's warning, and a cold sweat broke out on my forehead. My weapons lay on a low table only a few feet from me, yet I could not move to reach them without making sufficient noise to apprise my stealthy visitor of my whereabouts.

Another rustle, quite near me this time, was followed by the glow of a flashlight which swept the room, rested for a moment on my recumbent form, and then winked out. I sat up suddenly, at the sound of a scarbo drawn stealthily from its sheath not two feet from me.

No sooner had I sat up in bed than there was a whistling sound, followed by a thud, as the keen blade of a scarbo buried itself in the pillow where my head had lain a moment before.

I leaped from the sleeping shelf and fumbled for the light switch while my assailant, with a muttered exclamation of surprise and anger, flashed his torch on the coverlets. Then he whirled it around the room just as I found the switch and turned it.

Both of us were blinded for an instant by the glare of the light. I reached the table and secured my scarbo just in time to ward off his furious attack.

Back and forth we fought across the smooth floor, overturning furniture and tripping on rugs, while the apartment echoed and re–echoed with the clamor of our rapidly moving blades.

I found my assailant a dangerous antagonist; as a swordsman, Vorn Vangal was but a child compared with him. He was dressed in purple raiment trimmed with silver, and wore a heavy black beard.

At first his demeanor was one of sneering disdain; but when he found me able not only to parry his lightning cuts and thrusts, but to return them, measure for measure, a look of wonderment came to his hawk–like features. "Body of Thorth, stripling!" he exclaimed. "You have been practicing with the scarbo since I last saw you."

"I am but practicing now," I replied tauntingly, speaking slowly so that I might not mispronounce the words which came to me so readily.

His face reddened at this, and he redoubled his efforts, his keen blade flashing in shimmering arcs, alike bewildering and deadly. But his anger gave me the opportunity I sought. Whirling his blade on mine, as I had whirled that of Vangal some time before, I wrenched it from his hand and sent it clattering to the floor.

With a startled look he leaped back just in time to avoid a lunge that would have ended our conflict. As he sprang he shouted lustily, "Vinzeth! Maribo! Attend me!"

Two burly ruffians responded to his call, leaping through the window. They were armed with huge, broad–bladed spears and would probably have made quick work of me had not my own retainers burst through the door at my back, having heard the noise of our conflict.

For the moment the tide of battle turned in our favor. Then fresh re–enforcements poured in from outside. The leader had recovered his scarbo, and now they cut my men down until but a handful remained. Though our attackers were not without casualties, we were outnumbered from the start.

Maddened with the lust of battle, I was cutting my way through the spearmen in my endeavor to reach their leader when my tower guards made a sudden charge in response to a sharp order from their commander. At the same instant he plucked at my sleeve.

"The tower is lost, highness," he cried. "The traitors are too many for us. You must flee."

"Never! Let me at these assassins!"

I succeeded in breaking from his grasp, but he seized my arm once more, calling one of the guards to assist him. "Do not compel me to use force, Highness," he pleaded. "I must get you hence at once. To do otherwise would be treason to Your Imperial Sire."

The two of them dragged me through the doorway which they bolted. A moment later we entered the elevator and shot to the top floor, whence we climbed the spiral stairway to the roof. Far below us I heard the door crash inward—proof that the last guardsman had fallen.

They hustled me to the largest of the three airships, opened the door of the cab, and fairly hurled me onto the cushions.

"Raboth will take you to the palace," said the commandant. "I will bolt the door and follow in a one–man craft."

Raboth, a lean wiry youth with a thin, ragged beard, climbed in beside me and closed the door. As soon as he was seated, the ship began to rise—slowly at first, but rapidly gaining momentum until we shot upward with amazing rapidity.

My pilot, looking downward to take his bearings, drew back with a sudden intake of breath. "They have seen us! Two of their battle planes are rising to cut us off from the palace."

Scarcely had he spoken ere a searchlight flashed on our ship. An instant later a bullet ricocheted from our deck, tearing way part of the railing as it exploded. It had been fired from a mattork.

A terrific fusillade followed as we continued our rapid ascent. Suddenly we plunged into a thick cloudbank, shielding us from the revealing glare of the enemy searchlight. Continuing upward for several minutes more we cleared this lower cloud stratum and Raboth immediately put on our forward lights. Then he turned a switch, illuminating the interior of the cab with the radiance of a tiny bulb above our heads.

My pilot leaned forward to examine a small instrument suspended on a thin wire at the front of the cab. "I fear we are lost, Highness," he said, with a look of consternation. "One of the shells must have carried our magnet away. The compass is out of order."

A quick examination proved his statement correct. The magnet, which is fastened to the rear deck of all Olban airships to counteract the strong magnetic pull of the motive mechanism, had been snapped off by one of the mattork bullets. Now the needle pointed to the front of our craft no matter which way we turned.

A sudden glare of light at our backs, followed by the rending impact of a mattork shell on our hull, warned us that the enemy had sighted us. This time we dived into the stratum beneath us and then with level keel, hurtled forward at a pace that held me breathless with wonder.

"How fast are we traveling, Raboth?" I asked, trying to adjust my senses to the sight of cloud masses made iridescent by our lights, and moving past the cab in swift, bewildering kaleidoscopic display.

"This ship is rated at three–quarters of a rotation," he replied. "We are moving at top speed."

"What do you mean by three–quarters of a rotation?"

He seemed astonished at my question. "Why, a rotation is the speed at which Zarovia rotates on her axis. We are traveling three–fourths of that speed."

I made a rapid calculation. As the circumference of Venus is slightly less than that of Earth, and her day twenty–three hours and twenty–one minutes, Earth time, she rotates on her axis at a speed of more than a thousand miles an hour. Roughly, then, we were traveling at seven hundred and fifty miles an hour.

My companion held the ship to her course through the clouds for a considerable period, then dipped beneath them. This move almost resulted in our undoing; the second enemy craft, which had evidently been flying below us all the time, opened fire. I replied with our stern mattork— whether effectively or not, I could not tell—while Raboth again shot our craft up to the concealment afforded by the clouds. Once more we hurtled forward on a level keel.

"Our would–be assassins are certainly persistent," I remarked casually to my companion.

"And well they may be. This is the first time their leader has been recognized. No doubt we are the only two survivors of the fight in the tower, and consequently the only ones able to expose Taliboz."

"Who is this Taliboz?" I asked thoughtlessly.

"Is it possible that Your Highness does not remember Taliboz? He is the most powerful noble in Olba. For some time it has been hinted that he was conspiring against the throne, but there was no direct evidence. Now he must kill us all—both to do away with the heir to the throne, and to silence the witnesses of his perfidy."

We sped along for some time in silence. I calculated that if we had traveled in a reasonably straight line we were at least a thousand miles from our starting point. At length, feeling that we must have shaken our pursuers, Raboth once more descended beneath the lower stratum, taking the precaution of switching off all lights as he did so.

He looked about carefully, saw no sign of pursuit, and made the fatal mistake of turning on the lights. Scarcely had he done this ere a missile crashed through the back of the cab and exploded with a deafening noise. It struck on Raboth's side and killed him instantly, tearing his body to shreds.

Our lights were extinguished by the explosion, but a powerful searchlight played on us from behind and another shell carried away our stern. Then the craft lurched violently and fell, turning end–over–end while I clung desperately to my seat.

Chapter III

As THE wreck hurtled downward it gathered momentum each instant, and I expected nothing less than a terrific crash. To my surprise, however, the craft plunged nose first into water and sank rapidly. The cabin filled instantly through the great hole, torn by the mattork shell; but this same hole proved to be my salvation, for after the first cold shock of immersion was past I managed to scramble through it.

For several seconds I continued to sink in spite of my frantic efforts, due to the downward momentum of the craft I had just left. Then I stopped, and slowly began to make some progress upward, though it seemed at every stroke that my lungs must burst for want of oxygen.

After what seemed an age of lung–straining torture, my head bobbed above the surface, and I trod water while inhaling great breaths of the moist, salt air.

In the blackness of the Zarovian night, broken only at infrequent intervals by the momentary twinkle of a star or two through a rift in the ever–present cloud envelope overhead, I was unable to see in any direction. But I heard a familiar sound, far to my right—the roll of breakers on a windward shore. Toward this sound I swam slowly.

The sound grew louder as I progressed, and presently I lowered an exploring foot to find the bottom. Not reaching it, I swam onward once more. The second test proved more successful, and I stood erect, only to be knocked flat by a huge wave. I scrambled to my feet and, half wading, half swimming, at length dragged my weary body up on a sandy beach beyond reach of the breakers.

After a brief rest I arose and walked still farther inland, where I soon ran into a thick copse of bush–fern. The ground beneath the curved fronds was covered with moss, and on this I stretched, thankful for so soft a couch. In a short time, I was asleep.

I was awakened by the sound of voices quite near me. It was broad daylight and promised to be an exceptionally warm day. My silky scarlet garments had long since dried, as had my leather trappings, which had stiffened as a result of their soaking.

I judged from the tones that two people were conversing—a man and a girl. At first I did not hear what they said as I lay there on the soft moss only half awake, looking drowsily up through the rustling, wind–shaken fern leaves. Then the man raised his voice.

"Well you know, Cousin Loralie, that your parents desire the marriage as much as mine," he said in mincing patoa. "Is this not enough for you? Are you so lacking in respect for the wishes of your father and mother that you would set them aside for an idle whim?"

"Not for an idle whim, Cousin Gadrimel," replied the girl in a clear, musical voice. "I do not love you. What more need be said?"

"How do you know?" he demanded. "Yesterday we saw each other for the first time. We had but a few moments alone. I have not more than touched your hand. I could make you love me as I have…"

"As you have countless others, no doubt. Understand me, once and for all. No man can make me love him, nor could I make myself love any man, even if I desired to do so as a matter of filial duty."

Not wishing to play the part of an eavesdropper, however unintentional, I stood up, intending to offer my apologies and take my departure. As I did so I heard a muttered, "We'll see," from the man, followed by the sound of a struggle and a little scream of fear.

Pushing my way through the shrubbery, I came out on a moss–covered sward in the middle of which played an ornate fountain. Just beyond the fountain I saw a girl struggling to free herself from the embrace of a tall blond youth, whose yellow beard had just begun to grow. Both wore the scarlet of royalty.

"Let me go, you beast!" The girl's big brown eyes were flashing—her disheveled, dark brown ringlets flying as she struggled to free herself. Even in anger she was beautiful—more beautiful than any woman I have seen on three planets.

I sprang forward, seized the youth by the collar, and twisting it said, "If you are bent on wrestling this morning, Prince Gadrimel, permit me to offer you a more even match."

He released the girl and tried to turn, whereupon I twisted his collar the tighter. Then he reached for his tork, but I seized his wrist and bent it up behind his back. At this he began to bellow for the guard, whereupon I sent him crashing headfirst into the fern–brake.

I turned and bowed to the girl, who was still flushed and panting from her struggle. "Your Highness's pardon, if I intrude. It appeared to me that you were being annoyed."

"You were right, and I am indebted to you, Prince…?"

"Prince Zinlo of Olba," I finished for her, "at your service."

"I am the Princess Loralie of Tyrhana," she replied with a smile that revealed two adorable dimples. "Pray tell me…"

Our conversation was interrupted by the youth, who, after extricating himself from the bushes, rushed between us with drawn scarbo.

"Body and bones of Thorth," he snarled. "You have sealed your death warrant, Prince Zinlo."

Then he made a slash at me that would have severed my head from my body had I not leaped back. As I did so, I drew my own blade and engaged him. Finding in a moment that he was no master of fence, I disarmed him —then retrieved his weapon before he had time to recover from his amazement.

"You have dropped your scarbo," I said. "Permit me." And I presented it to him, hilt first.

Again he lunged at me, and again I disarmed him, with as much ease as before—then leaped and picked up his weapon before he could reach it.

"Perhaps I had better keep this," I said. "You seem so unfamiliar with its use that you may injure yourself."

He reached for his tork, but I was expecting this, and with a quick slash cut his belt. The weapon fell onto the soft moss, and I kicked it into the shrubbery.

He cringed as if expecting the death blow, then suddenly looked beyond me, exclaiming, "By the sixteen kingdoms of Reabon! Look behind you!"

Thinking it a trick, I did not look until I heard a scream from Princess Loralie and the clank of weapons. Then I whirled, and saw her struggling in the grip of a purple–clad noble whom I instantly recognized as my opponent of the tower—Taliboz! An Olban airship resting on the ground behind him explained his presence here. Four burly warriors were rushing toward me with drawn scarbos.

"It seems that we have some real fighting to do," I said to Gadrimel, tossing him his weapon. He caught it, and came manfully enough to guard, just as the four armed retainers of Taliboz bore down on us. I crouched low and extended my point as my first assailant made a vicious swing at my neck.

He died on my blade with an ear–piercing shriek, and I wrenched it free just as my second assailant came up. This fellow was not only more wary, but quite expert with the scarbo. He laid my cheek open with a quick cut just as I was coming on guard. His second blow was aimed at my legs, and would have mowed me down as grain is cut had I not leaped back. As it was, the point of his weapon raked my thigh.

Stung by the pain of my two wounds, I forgot my swordsmanship for the moment, and brought my blade straight down in a blow which he should have easily parried. It was the unexpected clumsiness of the stroke which told, as he did not come on guard in time; my blade divided his head as cleanly as a knife divides a Zarovian spore–pod.

Over at my left, Prince Gadrimel was sorely beset by the other two ruffians. His face and body were bloody as my own, yet he gave them back blow for blow and thrust for thrust. But he was plainly weakening. With the princess being carried off, there was no time for the niceties of dueling, and I felt no compunction about leaping up behind his nearest assailant and striking off his head. The other, seeing the blow, turned to face me; but to his own undoing, for he left Gadrimel the opening he sought. With a quick slash the prince disemboweled him.

"Come," I snapped, dashing toward the airship. "We must rescue the princess from that fiend."

He followed close at my heels, but we had not covered more than half the distance to the airship when it began to rise. Then a mattork projectile screamed past our heads, exploding in the shrubbery behind us, followed by another and another. We took shelter behind the marble rim of the fountain, and Taliboz's bombardment ceased.

The cannonading was suddenly resumed; but this time it came from the castle behind us. The castle guards, evidently believing themselves attacked by the Olban ship, were returning its fire with a vengeance.

Gadrimel and I both rose from our hiding place, and he shouted, "Don't shoot! The princess is on board."

The firing ceased, but too late, for the airship, its motive mechanism put out of commission by a mattork shell, was falling into the bay. I watched breathlessly as it hurtled downward, expecting to see it plunge beneath the water as my own had done the night before; but, to my astonishment, two parachutes flew upward from the fore and aft decks and effectively broke its fall. It alighted on an even keel with a great splash that nearly capsized a small sailing vessel anchored near by. Sinking no deeper than its deck railing, it rose again to ride the waves as evenly as if it had been built especially for the purpose.

Washed shoreward, it drifted closer and closer to the small sailing vessel while Gadrimel and I rushed down to the shore. Then, as we stood helplessly watching, a dozen armed men swarmed into the sailing vessel from the airship. The sailors instantly dived over the opposite side and swam for shore. The last man to step into the captured ship was the purple–clad Taliboz, carrying in his arms the limp form of Princess Loralie.

"To the docks!" shouted Gadrimel, racing madly off to the right. "They are raising the sails!"

As I hurried along, I saw the sails go up, billowing in the breeze, while four of Taliboz's men at the prow hoisted the anchor.

Gadrimel and I rounded a bend in the wooded shore line, and a crescent of docks to which several hundred ships were moored came into view. At the same time, the vessel which Taliboz had captured, with all sails up and anchor hoisted, veered about in the considerable breeze and made swiftly for the open sea.

A party of soldiers from the castle had reached the dock ahead of us. With them was a tall, broad-shouldered figure in the scarlet of royalty, whose grizzled beard was cut off square below the chin, and whose regal countenance was empurpled with anger.

"It's my father, Emperor Aardvan of Adonijar," said Gadrimel.

"Prepare six warships for pursuit, at once," I heard Aardvan shout.

A thousand men hurried to carry out his orders.

As we approached this commanding individual, the prince and I both bowed low, with right hands extended palm downward, in the universal Zarovian salute to royalty. I was struck by the contrast between this brawny, bull-necked emperor and his mincing, effeminate son.

Aardvan, glaring down at us, roared, "Two brawling princelings, all spattered with blood. What did you do? Scratch each other like a couple of marmelot cubs? Who is your playmate, Gadrimel? Were those his men who carried off the princess?"

"This is Torrogi Zinlo of Olba, Your Majesty," replied Gadrimel.

"The Imperial Crown Prince of Olba! What does he here?"

I explained briefly.

"We slew four men, sire," boasted Gadrimel.

"I've heard of this Taliboz," growled Aardvan. "A traitorous and dangerous fellow. You are welcome to Adonijar, Prince Zinlo. Stay as long as you like, and when you are ready to depart I'll send a guard of honor to accompany you to your own country."

"With your majesty's permission," I said, "I should prefer to accompany the fleet which is preparing to follow Taliboz."

"That will be as Gadrimel says," rumbled his father. "He will command the fleet."

"Come along," said Gadrimel. "Our private quarrel can wait. For the present we have common interests, and your blade may be needed."

A gray–bearded naval officer came running up and saluted.

"What is it, Rogvoz?" inquired Emperor Aardvan.

"The fleet is ready, Your Majesty," replied the officer.

"Then let's be off," said Gadrimel.

We hurried aboard one of the six vessels, all of which swarmed with armed men, accompanied by the gray–bearded officer. A few moments later, with all sails set, the fleet plowed out of the harbor in pursuit of the small fishing boat, which was now but a speck on the horizon.

Chapter IV

The tiny sailboat in which my mortal enemy, Taliboz, was carrying off the Princess Loralie, was making steadily northeast toward Olba with our six battleships in hot pursuit, when suddenly I saw her come about and head directly south.

Gadrimel, Admiral Rogvoz and I were watching together on the forward deck of the flagship. The admiral stared for a minute through his long glass. Then he carefully scanned the horizon toward the northeast.

"They have good reasons for turning," he announced excitedly. "A great ordzook approaches from the north!"

He passed the glass to Gadrimel, who looked for a moment, then with an exclamation of horror, passed it on to me.

When I had adjusted the glass to suit my vision, I saw a most fearsome sight. Not more than a half mile behind the small sailboat, and gaining on it rapidly, a gigantic and terrible head projected from the water, swinging on a thick arched neck. The head alone was half as long as the sailboat it pursued; and although the body was submerged, I could see, at intervals of fifteen to twenty feet, sharp spines flashing intermittently above the waves to a distance of fully a hundred feet behind the head.

"Do you think we can save them, Rogvoz?" asked Gadrimel.

"We can but try, Highness," replied the admiral. "It is doubtful." He turned to the captain of the boat. "Order the mattork crews to start firing on the ordzook, and signal all other captains to do likewise."

The captain shouted his orders to the waiting cannon crews, and a moment later the din of these rapid–fire weapons was terrific. From the high forward deck our signal man meanwhile busied himself semaphoring with two huge disks, one red the other yellow. The other ships immediately opened fire with their mattorks, adding to the deafening noise which our own ship had started.

We were approaching closer to the marine monster now, as the path of the fishing boat crossed our own. I could see the ordzook turn from time to time, snapping at the stinging mattork projectiles as they struck the spiny ridges of its undulating scaly body, which was a shimmering, bluish–green in color. The head and neck were a brilliant shade of yellow, except where neck and shoulders joined, for at this point a broad band of scarlet formed a flaming ring—a danger signal which all creatures might beware.

The speed of the mighty amphibian was impeded by its constant turning to snap at its wounds, enabling the small boat containing Taliboz and Loralie to gain on it gradually.

Suddenly changing its course, the monster wheeled and swam toward our fleet. "To the right!" called Rogvoz. "Veer to the right!"

The ship on which we stood came about suddenly, her starboard rail for a moment submerged beneath the waves. All hands grabbed for such fixed objects as they could cling to.

Behind us trailed the fleet, and on came the ordzook, not stopping now to snap futilely at the stinging projectiles, but bent on more deadly action.

With all the port mattorks trained on the monster, I thought to see it go into a death struggle at any moment, but the projectiles seemed merely to irritate it. We were so close in a few moments that I could see its relatively tiny jet black eyes, set just above the corners of the great gaping mouth which was filled with a formidable array of saw–edged teeth.

We passed it safely, as did the second, third, fourth and fifth boats, but the last of the fleet, lagging behind because of improper manipulation of its sails, could not escape.

The enormous yellow head reared upward for an instant on the arched, spiny neck. Then, with formidable jaws distended, it struck downward at the fore deck. The captain of the ship and three of his men standing with him disappeared into the huge maw along with most of the deck on which they stood.

Again and again the creature struck at the doomed craft, until sails, masts, men, and most of the upper works were gone. Then it reared upward in the water and came down with a tremendous crash on the middle of the defenseless hulk. Broken in two by the terrific impact, both halves of the ship sank almost instantly, and the fearful creature which had wrought this destruction before our eyes plunged into the waves after them. Nor did we see it more.

Once more we turned our attention to the boat containing Taliboz and the princess. Hemmed off from Olba by our five vessels, they were now sailing due south at a speed apparently equaling our own, for as time passed the distance between us did not seem appreciably to alter.

Because of the presence of Princess Loralie on board the fishing boat we were constrained to withhold our mattork fire, with which otherwise we could soon have brought Taliboz to terms. He fired no shots, either, except a few stray projectiles from the torks, which led us to believe that he had not salvaged any mattorks from his wrecked airship.

As we sailed southward over the blue–gray waters of the Ropok Ocean, the point of land on which the city of Adonijar is situated receded from view, and in all directions showed only a cloud–lined sky meeting and almost blending with the rolling waters.

But even this vast expanse of sky and sea was not a lonely place. It teemed with life of a thousand varieties—with creatures of striking beauty and of the most terrifying ugliness. Quite near our boat several large white birds with red–tipped wings and long, sharply curved beaks skimmed the water in search of food. Mighty flying reptiles, some with wingspreads of more than sixty feet, soared high in the air, scanning the water until they saw such prey as suited them; then, folding their webbed wings, they plunged with terrific speed, to emerge with struggling prey and leisurely flap away.

With the advent of sudden darkness, common to tropical and semi–tropical Zarovia, bright searchlights flashed out from the mast–heads of the entire fleet, and the boat we pursued was thus kept in sight.

While these lights were an absolute necessity in the blackness of the moonless Zarovian night, they were also a nuisance, as they attracted to the vessel countless droves of flying creatures, mostly reptilian; many of them, blinded by the bright beams, flew against masts, sails or rigging and fell, squawking, croaking or hissing to the deck. Some of them, infuriated and only partly crippled or stunned, menaced our lives until dispatched and tossed overboard.

After several hours I grew weary and retired for the night. Despite the constantly repeated disturbances above deck and the frequent colliding of the craft with some marine monster, I soon fell asleep.

I was awakened late the following morning by Prince Gadrimel's valet, who insisted on ministering to my wants as became a prince of the blood imperial. After a breakfast of stewed mushrooms and succulent grilled fish, washed down with a bowl of steaming kova, I went on deck where I found Gadrimel and Rogvoz in consultation.

"They swing gradually but surely toward the southwest, Highness," said Rogvoz as I came up. "They are trying to circle us and sail once more toward Olba."

"Is there no way we can prevent their doing this?" asked Gadrimel.

"We can only follow them so that their circle must be so large that they will be cornered by land."

I took up the glass which he had put down in order to make some calculations, and focused it on the ship we were pursuing. On the rear deck I made out the slim figure of the princess, who also held a telescope in her hand. She raised it a moment later, and I saw that it was pointed at our ship. I waved my left arm.

Her reply was instantaneous, as her shapely white arm flashed above her head. Then I saw Taliboz, glowering with rage, come up behind her, wrench the glass from her grasp and with significant gestures order her forward. With little head held high, she defied him, but he grasped her wrist and dragged her away. As she disappeared from view, I lowered my glass, and Gadrimel, who had evidently been watching me, said, "Beard of Thorth, Prince Zinlo! Your usually serene and smiling countenance has suddenly become as stormy and forbidding as the Azpok at change of seasons. What have you seen?"

"Enough," I replied, "to make me long for the day when I can once more meet Taliboz face–to–face, scarbo in hand!"

For five days we followed in tormenting nearness, sometimes close enough to be within hailing distance, sometimes so far back that we feared to lose them. It was late on the fifth day that a lookout at the masthead above us suddenly shouted: "Land! Land!"

Instantly Gadrimel, Rogvoz and I rushed to the foredeck. Taliboz, now hemmed in from all sides by our fleet, was doing the only thing left for him to do, steering directly for a sheltered inlet. He rounded a curve in the shore line, disappearing from view, and some time later, when we sailed into the inlet, we saw his craft beached.

Rogvoz, who had the glass, exclaimed, "The fool! The utter fool! To escape us he plunges into worse danger, dragging the princess with him. We, at least, would not eat him."

"What do you mean?" demanded Gadrimel.

"Just now I saw the entire party disappear into the fern forest."

"But this danger you mention. What is it?"

"I had forgotten, Highness, that you are unfamiliar with this part of Zarovia. This is the land of the terrible, flesh–eating cave–apes—huge creatures, any one of which is said to be a match for a dozen men, but with intelligence far greater than that of other apes. Some of the few men who have landed here and had the good fortune to escape them say they not only have a peculiar clucking language of their own but can also speak patoa."

"We must catch up to them quickly," I cried.

The five ships were brought up as close to the sloping, sandy beach as was safe, then boats were lowered. Soon a force of five hundred fighting men stood on shore.

After a short consultation, it was decided that we should form a long line, the men keeping about ten feet apart, and so enter the forest in the direction which Taliboz had taken. This line, if kept unbroken, would form a great net nearly a mile across in which the fugitives, we felt sure, must inevitably be snared. Rogvoz took charge of the extreme left end of the line, Gadrimel directed the center, and I had charge of the extreme right end.

Tripping over clinging creepers, floundering through sticky morasses, cutting our way through matted, tangled ropelike vines which hung downward from the mighty branches of the tree–ferns, and constantly slapping at the biting and stinging insect pests which abounded in these lowlands, we soon found ourselves progressing with exasperating slowness.

Not only did the vegetable and insect world seek to detain us, there was the menace of animals and reptiles as well. A giant whistling serpent—a hideous creature fully forty feet in length, with long, upright ears and sharp spines the full length of its back—struck down one of our men and succeeded in killing two others before it was finally dispatched by the bullets from a score of torks.

Soon the men had banded in groups of about twenty each for mutual protection.

The group nearest us lost three men to a ramph, a great hairless bearlike creature, whose scaly hide was a brilliant chlorophyl green above, fading to a greenish yellow below. After they had slain it they fell to with their scarbos, cutting it up and bearing portions of the meat with them, for ramph steaks were considered the most delicious meat on Zarovia.

Some time near noon, my party was attacked by a marmelot, a vicious feline fully as large as a terrestrial draft horse, its hairless, scaly hide a mottled orange and black, its great saber tusks fully a foot in length. Seven of our men were slain by this, one of the fiercest of the Zarovian jungle creatures, before it was dispatched.

Brave men were these soldiers of Adonijar; in spite of the sudden death which hovered over us in these tangled jungles, they cut their way forward without grumbling or word of turning back.

Because they had stopped to cut up the ramph they had slain, we had lost sight of the party next to us, and it was not until darkness suddenly descended that I thought to communicate with them. I called out to them then to halt, but received no reply. Again I called at the top of my voice, but there was no answer.

"Remain here," I told my men, "and I will go and find them. They cannot be far away."

Glad for a rest after their arduous march, the group quickly cleared a place for a fire, and got out their kova and provisions to prepare their evening meal.

I then set out in the direction which I felt sure would lead me to the next group of warriors, flashing my light ahead of me. I must have traveled for at least two miles, shouting from time to time without receiving any reply, when suddenly I heard a quavering, mournful howl from the darkness at my right.

Swinging my light around in the direction of the noise, I saw three huge, slinking forms and three pairs of blazing eyes. They slightly resembled terrestrial wolves, but were fully twice as large as any wolves that ever lived on Earth. Their scaly hides were slate gray in color, and each had a ruff of long, sharp spines which stood out around the neck like a spiked collar. Upon describing them later, I learned that they were awoos—so called, no doubt, because of their doleful, nerve-racking cries.

Swinging my tork into line, I instantly brought down the foremost beast, whereupon the others crouched, disappearing from view. Howl after howl resounded from all directions. They began to close in on me.

I whirled this way and that, and where the light was caught by the glowing eyes of the wary creatures, my tork spat death, but I soon saw that it was a hopeless fight. It seemed that as soon as I killed or wounded one creature, two more stepped in to take its place.

There was nothing left for me to do but to climb into the branches above me, hoping they would be unable to follow. Accordingly I swarmed up one of the trailing, rope-like vines which hung from the mighty fronds of a tree-fern fully sixty feet above my head, and soon found myself in a huge leaf crown which afforded a temporary resting place.

The howling chorus below was terrible to hear. The pack, now more than a hundred in number, milled about the base of the tree while the more impatient of the creatures leaped up, snapping and snarling. Time and again I used my tork, littering the ground with their carcasses, but the dead brutes were instantly replaced by others.

Wondering how long this sort of thing would last, I was slipping a fresh clip of gas and one of projectiles into my weapon when I heard a rustle of the leaves above me. Glancing upward, I beheld a huge gorilla–like face surmounting a mighty chest fully three feet across. Then a great hairy hand descended on my head with terrific force, and I lost consciousness.

Chapter V

When I had once more become aware of my surroundings, I was lying in semi–darkness on a cold stone floor. The top of my head was bruised and tender, and my neck so lame that a sharp twinge of pain shot through it each time I turned my head to look about. The belt, to which my tork and scarbo had been fastened, was gone.

I sat up, and my brain swam dizzily for a moment. My vision cleared presently, and I saw that the source of the light which but faintly illuminated the spot I occupied was a jagged opening—evidently the mouth of a huge cave.

Quite close to me on my left, I became aware that some creature was breathing heavily, apparently in sleep. Turning, I beheld the recumbent form of a gigantic hairy female—head pillowed on arm, and knees drawn up as if for warmth, sleeping not four feet from me.

The face was neither ape nor human, but partook of the characteristics of both. The form, slender of waist, full–breasted and broad–hipped, was more like that of a human female than a she–ape, though covered with short, reddish–brown hair. The limbs were not ungraceful, but the toes were long and evidently prehensile. I judged that the creature, when standing erect, must be at least eight feet in height and so powerfully muscled as to be a formidable antagonist.

Stealthily I stood erect, then tiptoed toward the mouth of the cave. I had not taken more than a dozen steps when something tripped me and I fell headlong to the jagged floor. At the same time there came the sound of a fearful growl behind me.

Before I could scramble to my feet I was pounced upon from behind and jerked erect. Then, with my arms pinioned behind me by two powerful hairy hands, I was marched out into the sunlight. Looking up, to the considerable inconvenience of my injured neck, I saw that my captor was the big female who had been sleeping so peacefully a moment before. She had been awakened by a thin but exceedingly tough twisted string of gut, tied to my ankle and her wrist.

We were high up on a rugged hillside which seemed honeycombed with caves. In the valley far below us, I saw the waving fronds of huge tree-ferns above the tangled mass of jungle vegetation.

"So, food-man, you would escape Chixa, and thus have Chixa slain," said my captor in a peculiar, clucking patoa.

"It is high time you were taken before Rorg. Perhaps he is hungry."

"Release my wrists," I replied, "and I'll be glad to go with you before Rorg. Who is he, and what has his hunger to do with me?"

"Rorg is the king, the Rogo of the Cave-Apes." The tall female released my wrists and stepped up beside me, taking a firm grip on my right arm. "If he is hungry he may want to eat you."

"What makes you think I will be good to eat?" I asked.

"I have tasted the flesh of many food-men, and most of it is good, though it is sometimes too salty. Are you very salty?"

"Very. I'm afraid your ruler would be displeased."

"If you are very salty he will be greatly pleased," said Chixa. "He likes salty food-men, though I do not."

About the furry waist of my captor there was a string like the one bound to my ankle. Swinging from this string on the side opposite me, by a short hook in the handle, was a weapon I greatly coveted.

It was a club of hard wood about three feet in length, shaped something like the blade of an oar, but thicker and heavier, and pointed at the end. Set in the two edges of this club were small bits of sharp flint which gave it a formidable saw-like appearance. It was heavy enough to crush a skull or break a limb, and sharp enough to lacerate the toughest muscle. A large flint knife also swung between her breasts from a cord around her neck.

The cave-ape walking beside me was in some ways like a woman, and because of that faint similarity I hesitated for a moment to carry out the plan which had come to me. But life has ever been dear to me—even though I love adventure so greatly that I have risked death in many terrible forms on three planets—so my hesitation was but momentary.

Suddenly turning with my right arm bent at the elbow, I put all my weight in a blow that landed in the furry solar plexus. With a terrible sound—half scream, half roar—my tall captor clasped her hands to her abdomen and bent over. As she did so I pivoted the other way with a left to the point of her jaw, and she fell unconscious at my feet.

Quickly slipping the knife cord from around her neck, I sawed the gut tether from my ankle. Then I seized the club which dangled from her belt, and looked about me for the most likely avenue of escape.

To my surprise and horror, I saw that there was none, for at the sound of Chixa's voice, the caves had suddenly spewed forth not less than a thousand of these gigantic creatures, all armed as I now was, with flint knives and sawedged clubs. The mature females varied in height from seven to nine feet and the males from ten to twelve.

Those nearest me had spied me as I got to my feet, and now approached menacingly from all sides with bared fangs and low, throaty growls—the males displaying long, downcurving tusks which greatly increased their ferocious appearance.

With the club held swordlike in my right hand, and the flint knife gripped in my left, I leaped for a great leaning boulder, one side of which could afford me protection from above and behind.

A huge tusked male sprang forward to bar my progress, and swung his saw–edged club in a terrific blow. He was fully eleven feet in height, and towering above me as he did, offered no opportunity for quick club work.

There was, however, a chance to use the knife, which I did without compunction. Leaping beneath his swinging arms, I buried it in the right side of his abdomen and ripped him across the belly. While he swayed drunkenly, I completed my rush to the temporary protection of the boulder, and as I turned with my back against it to meet the attack of the others, I saw him topple to the ground.

A moment later I was confronted by a semicircle of growling, roaring cave–apes, swinging their clubs menacingly, but a little different about approaching me too closely—probably because of what had happened to their companion. Mixed with the growling and roaring I could distinctly hear the patoan words "kill" and "meat," which sounded ominous enough.

The great tusked males seemed to be working themselves into a frenzy of fury as they came closer and closer—evidently their primitive way of attempting to overcome their fear of me.

Presently one leaped out ahead of the closing line, and swung his club for my head with a terrific downward, twohanded stroke. I stepped to the left, and forward, and as his club was shattered on the stone where I had been standing, the flinty edge of my own bit deeply into his cervical vertebrae. He fell on his face without a sound.

I sprang to a new position, brandishing my club menacingly, and the line of attackers moved back a little.

"Kill! Kill!" The word was repeated constantly now as the savage semicircle closed in once more.

"Come and be killed!" I replied.

"You will be next to die, food–man," roared a huge male who stood near the center of the line, "for Urg is about to kill you." Urg stood at least twelve feet in height, a head taller than the other males in the front line, and his great downcurving tusks, fully seven inches in length, gave him a most ferocious aspect.

He seemed about to spring forward, and I had braced myself for his attack, when there was a sudden commotion behind him. The milling crowd of apes drew back respectfully to make way for a huge male, taller and heavier even than Urg.

Just behind him walked two young females, one waving a fern frond to keep annoying insects away from him, while the other carried a huge gourd–like fungus with a bottle neck and a bowl made from a split sporepod. Behind these two walked more ape–maidens, some carrying fresh meat, while others bore bowls heaped high with fragments of edible fungi or sporepods, cracked, and ready for eating.

Coming up behind Urg, the newcomer carelessly pushed him aside and stood in the front line, surveying me with apparent boredom. At this, Urg gave a low growl, whereupon the larger ape smote him in the mouth.

"Growl again at Rorg, and you will feel the weight of his club."

"I did not know it was Rorg who pushed me," replied Urg.

"Why do you hesitate before this little food–man?" asked Rorg. "Do you fear him?"

"Of course not," answered Urg. "I was playing with him. I was about to kill him when you came up."

"I believe you fear him," continued Rorg. "I notice he slew your brother, Arg, who was as good a fighter as you. This is unusual for a food–man. He must be a mighty warrior among his people. It shall be for Rorg, mightiest of the cave–apes, to slay him."

"It is my right to kill him," growled Urg, "for he slew my brother."

"He will be killed when and how I ordain, for I am king." He swung on me once more. "Who are you, food–man," he asked, "and how did you slay my people?"

"I am Zinlo," I replied, "and I slew your people with the weapons of Chixa which I took from her."

"How could you take Chixa's weapons from her?" asked Rorg incredulously. "Why, she is ten times as strong as you. I do not believe it. Chixa gave you her weapons, so Chixa shall be slain."

"Chixa lies unconscious on the ground, Rorg," clucked a female. "This food–man must have taken her weapons by force."

"Chixa is feigning and shall be slain," said Rorg. "Such a thing would not be possible. Go and slay her, Urg."

All this time I had been standing guardedly, saying nothing; but when it became apparent that the female ape was about to be killed through no fault of her own, but because of something I had done, I felt a wave of pity for her. Brute and man–eater though she was, she had been considerate of me. After all, she was something like a woman.

"Rorg," I said, "I did not lie about taking her weapons from her, and I can prove it."

"How?"

"By taking the weapons from your strongest warrior in the same manner."

"Can you take Urg's weapons from him?" asked Rorg.

"Of course."

"Then you must be very strong or very clever. I like clever food–men. Sometimes I keep them for a long while when they are exceedingly clever. When they fail to amuse me they die. Let me see you take Urg's weapons, and I will spare your life for today, at least."

"But what of Chixa?"

"I will spare her life, also."

"Good. I will need plenty of room, and I demand your promise that I will not be attacked by any one other than Urg."

"You will have plenty of room, and you have my word that you will not be attacked or interfered with," said Rorg.

"Move back, then, all of you," I said, "until I tell you to stop."

The crowd drew back until the front line was a hundred feet from the rock in all directions.

"That is enough. Now, Urg, come here and I will take your weapons. I will go unarmed, and you must not have your weapons in your hands. You will walk beside me as if I were your prisoner fastened to a tether." With this I dropped weapons to the ground.

"It is a trick," growled Urg, but at Rorg's command he hung his flint knife around his neck, and hooked his club in the string around his waist. As the brute lumbered up beside me, and I saw what a mighty tower of strength he was, I must confess that I felt considerable doubt about being able to knock him out.

He strode along beside me, his great arms swinging at his sides. I timed my swing for the instant when the great paw nearest me was back, leaving the abdomen unguarded. Then I pivoted, landing my right fist in his solar plexus—all the force I could muster behind it.

With a grunt of surprise, he doubled forward as Chixa had done; but before I could swing for his jaw, he stood erect once more and reached for his club. His chin, by this time, was so high in the air that I could not reach it, and he had his plexus covered by his great forearm; there was nothing I could do with my fists. His shins; however, were exposed; I kicked the right one with my sandaled foot.

Uttering a howl of pain, he raised his foot and launched it at me, whereupon I grasped it with both hands, and twisting it with a sudden jerk that caused the bones to creak, turned his toes downward and his heel upward at the same time. This turned him completely around, and a quick push sent him on his face.

Before he could scramble erect, I leaped on his back, planting a heavy blow just beneath his ear. He shook himself in an effort to dislodge me, but I grasped one of his tusks with my left hand, and with my legs wrapped around him, continued to hammer him behind the furry ear.

Standing erect, he bellowed angrily, and releasing his grip on his club, grasped my left arm in his huge right hand. Wrenching my hand away from his tusk, he jerked me forward over his left shoulder and threw me to the ground fully twenty feet away. Fortunately for me, I alighted on my feet, and although I stumbled and fell, was unhurt.

I saw Urg coming toward me, but he reeled drunkenly.

Quickly springing to my feet, I leaped forward, whereupon he jerked his club from his belt and made a wild swing for my head. As his momentum bent him forward, I dodged, and leaping in, planted a blow in his right eye. He straightened, and I struck him in the solar plexus once more.

This time he doubled up, exposing his jaw, on which I planted a crashing right hook. Once more he stood erect, tottering unsteadily, and once more I doubled him up with a plexus blow, getting in a left to the jaw. He fell on his face as I sprang out of his way, finishing him with a blow behind the ear.

I slipped the knife cord from around his neck, and picked up the great saw-edged club which he had dropped. Then I leaped upon his back, and with one foot on his neck, brandished the weapons aloft, while a great howl went up from the mob around me.

From his place in the center of the line, Rorg walked slowly toward me, attended only by the female with the fern frond. I stepped down from the prostrate body of Urg as he approached, and slung the knife about my neck, also hooking the club in my belt. "Are you convinced?" I asked.

"I am convinced," replied Rorg. "You are clever enough to be kept alive for a while, and Chixa shall be spared."

It was then I noticed a gold bangle about Rorg's wrist. I saw that it was stamped with the coat of arms of Taliboz, and it followed that this must have belonged to one of his retainers.

"Where did you get the man who wore that bangle?" I asked.

"My warriors captured him with twelve other food-men, and a food-woman. We have eaten them all, except one man who is very clever, and the woman, who is very beautiful."

"Do you know the name of this clever food-man?" I asked.

"His servants called him Lord Taliboz," was the reply.

"And the food-woman?"

"A royal princess, fit only for royalty. I intend to wed her at the beginning of the next endir. Although I should like to wed her sooner, I will not depart from the customs and traditions of my forefathers, who married but one wife at a time and her at the beginning of each endir, thus taking but ten mates a year. I had intended Chixa for my next wife, but now she will have to wait for another endir."

"Is it customary for cave-apes to mate with food-people?"

"It is not," replied Rorg, "but we have no old law against it. I make all the new laws, and I have decreed that, hereafter, all Rogos of the Cave-Apes may marry food-women if they choose to do so."

"I have a great curiosity to see this food–man who is so clever and this beautiful food–woman," I said.

"You shall see them," replied Rorg. "Come with me. I want you to do some more clever tricks, anyway, to amuse my wives and children."

As I strolled away with Rorg I saw Urg stir slightly, then roll over and sit up, after which he tenderly felt his bruised jaw and the battered spot behind his ear.

Chapter VI

Rorg and I climbed high up the mountainside while his female attendants and the mob of cave–apes which had been so bent on killing me scrambled after us.

We were ascending the tallest peak of a chain of mountains which extended toward the north and south, their rugged slopes partly concealed by the various strata of gray clouds which floated lazily westward. And these mountains, as far as I could see, swarmed with cave–apes.

As we mounted steadily upward we passed many ape families, some of which were breakfasting while others appeared to be starting out on their morning quest for food. Tiny helpless infant apes were at their mothers' breasts. Spindle–legged, round–bellied ape children played about on the rocky slopes, or gnawed at bones, scraps of meat, edible fungi, and sporepods.

All of them, from babes to adults, watched me with their beady black eyes as I passed, but none made a hostile move or sound, evidently because of the awesome presence of Rorg.

At length we climbed over the rim of what had once been an active volcanic crater. It was shallow, filled with the litter of centuries. In the center a volcanic cone projected upward, and toward this we made our way across the debris–strewn crater floor. The walls of the crater, I noticed, were honeycombed with caves.

Enormous male apes, some of them nearly as large as Rorg, patrolled the rim of the crater, their saw–edged clubs swinging in their hairy paws. With these alert sentries always on duty, it was plain that escape from the crater would be most difficult and dangerous.

As we drew near to the mouth of the great central cave a number of females and young ones of assorted ages and sizes came out.

"These are my wives and children," said Rorg. "If you are as clever as I think you are, you will find a way to amuse them."

"I will find a way," I promised, "but first let me see this clever food–man and beautiful food–woman of whom you have told me."

"I will send for them at once."

Searching in the debris near the cave mouth, I picked up two well–dried finger bones which looked exactly alike. Palming one and displaying the other as I stood with my face to the audience and my back to the wall of the volcanic cone, I proceeded to perform some very simple tricks, such as making a finger bone disappear from my right hand—then seemingly plucking the same finger bone out of my ear with my left. I even appeared to remove six finger bones, one after another, from the ear of one of Rorg's half–grown male children.

My audience seemed intensely interested in what I was doing, but I noticed that no matter what tricks I performed, not one of them laughed. Then I remembered that, to them, I was actually doing the things I seemed to do.

Before I had performed many tricks I saw two figures coming toward me, each tethered by the ankle to the wrist of an enormous she–ape. Instantly, I recognized the purple–clad, black–bearded Taliboz, and the slender, scarlet–draped figure of Loralie.

Rorg, who had seated himself on a low boulder with his female attendants behind him, ordered Loralie to a place on his right and Taliboz on his left.

With right hand extended palm downward, I bowed low to the princess in the customary salute to royalty, but she did not respond, nor even give any indication that she had seen me. Instead, with a haughty toss of her pretty little head, she sat down at Rorg's right and, looking across at Taliboz, said something in a low voice which I could not quite catch. He smiled unpleasantly at me.

Puzzled at this singular and inexplicable show of dislike on the part of the princess, I mechanically went through several more tricks from the book of magic—then pocketed my bones and bowed.

"You are indeed clever, food–man," said Rorg. "You are even more clever than Taliboz. To pluck six bones from the ear of Vork! I will not eat you today. You may go now, without tether or guard, but do not attempt to pass the crater rim or you will die."

I walked away with the black beady eyes of the cave–apes staring after me and the sardonic grin of Taliboz following me. But Princess Loralie deliberately looked in another direction.

As I wandered about the crater I pondered the strange conduct of the princess. What could I have done—or what could Taliboz have told her—to arouse her anger and disdain to such a degree that she showed it even when we were both in deadly peril and should have united forces against a common enemy?

I was half oblivious of my surroundings until a hairy paw was laid heavily on my shoulder. Quickly whirling, I faced a huge ape about eleven feet in height whose scarred fur was spotted with gray, attesting his considerable age.

"I am Graak," he said. "Rorg sent me to feed you. I have food in my cave. Come."

The old warrior turned and I followed him across the crater past many ape families, who looked at me curiously, but manifested no special hostility. Presently we came to a rather small cave, the floor of which was littered with old and malodorous gnawed bones. From the partly devoured body of a huge ptang, or giant sloth with sharp upcurved claws, he carved a slice of raw meat which he handed me.

"I slew the ptang this morning," he said, "so it is fresh and good."

Casting about for fuel, I found a pile of dried fern fronds near the entrance. After powdering a quantity of them, I at length succeeded in igniting them by striking my flint knife against one of the buckles of my leather trappings, and soon had a small cooking fire crackling. Over this I held my ptang steak impaled on a fern frond.

Graak watched me with evident wonder. "You are indeed a sorcerer."

For three days and nights I ate the food which Graak brought me and slept in his cave. Although his manner was surly, he was never openly hostile.

But all my attempts at cultivating his friendship failed.

I spend most of my daylight hours searching for the cave in which the princess was confined, but it was not until the morning of the fourth day that I found her, seated in the doorway of a cave quite near my own. She must have been purposely avoiding me.

I swallowed my injured pride, and stepping before her, bowed with right hand extended, palm downward. "Prince Zinlo craves a word with Her Highness, Princess Loralie."

She did not answer, but turning her head away as if she had not heard me, addressed something to her huge female guardian.

Without moving, I repeated my request.

She rose with flashing eyes. "Begone!" There was withering scorn in the look she gave me. "Annoy me further and I will call the apes and have you driven away."

I bowed and departed. There was nothing else left for me to do.

Just before I reached Graak's cave, I came face-to-face with Taliboz, walking with his huge female guard. He grinned maliciously and said, "Tomorrow is the first day of the fourth endir."

"Any fool knows that," I retorted.

"Perhaps any fool also knows that on the first day of each endir, Rorg takes a mate. And that if food-men are available, a food-man is served at the wedding feast." As I stared at him, he added, "Rorg has just promised me that I shall not be eaten tomorrow."

I sat down before Graak's smelly cave. On the morrow, Rorg was to take Princess Loralie as his mate, and there were but two food-men held prisoner by the cave-apes—Taliboz and myself.

As we breakfasted on fungi and sporepods the following day, Graak was more talkative than he had yet been. "Today is Rorg's mating day with the food-woman—if he lives," he said.

"What do you mean?"

"Some of our bravest warriors do not want our race to degenerate by intermarriage with weaklings. There has been much talk, and I believe Rorg will be challenged."

"Who will challenge the king?"

"It is the privilege of any warrior to challenge the Rogo to a duel to the death on the mating–day. The warrior who succeeds in killing him becomes Rogo in his stead, and takes his prospective bride as well as his other wives, children and possessions."

"But suppose one of your warriors who does not believe as Rorg believes slays him. What will then become of the food–woman?"

"She will be eaten, and Chixa, who was cheated of her turn, will be taken as a mate."

As Graak and I finished our meal, I noticed that the crater was beginning to fill with apes. Young and old, male and female, they came at first in scattered family groups, but later in great droves, until the huge pit was literally seething with moving brown figures.

Presently a tall, yellow–tusked male shouldered his way through the crowd and stopped at the door of our cave.

"Rorg commands the presence of Zinlo, the food–man," he said.

As I trailed the huge ape through the jostling throng, I tried to formulate some plan of action by which the princess might be saved. Although I resented her attitude toward me, I felt the urge to fight in her defense.

We came at length to the mouth of Rorg's lair in the great central cone. Passing through the deserted cave, dimly illuminated by reflected light from the exterior, we stepped into a narrow runway which slanted upward in a long curving spiral.

As we progressed steadily upward, the way grew so dark that I was forced to hold out both hands to avoid running against the walls. Presently it became lighter once more, and in a few moments we emerged onto the flat, narrow top of the cone.

Squatting in a semicircle near one edge of the platform were a dozen cave-ape warriors. At one end of the semicircle I recognized Urg, the huge ape I had disarmed, leaning on his great, saw-edged club and looking as belligerent as before.

Near the rim just opposite this ring of warriors stood Taliboz and Princess Loralie. Although their huge female guards stood behind them, I noticed that their tethers had been removed. The traitorous Olban noble favored me with a leer as I emerged from the runway, but the princess would not so much as notice my coming.

In the very center stood Rorg, evidently awaiting my arrival as he looked down at the vast sea of upturned faces in the crater. I was placed with my back to the twelve warriors.

As soon as I had taken my position, Rorg held his sawedged club aloft. Instantly the vast murmur of voices from below was stilled.

"Your Rogo takes a mate," he bellowed, his deep tones reverberating from the surrounding crater walls. Then he leaped high in the air, brandishing his saw-edged club until the air sang and whistled through its teeth. Alighting with a loud smack of his leathery feet on the hard rock, so that he faced in a direction opposite to that in which he had previously looked, he roared once more, "Your Rogo takes a mate." Leaping, whirling, and alighting as he had done before, he made his announcement in four directions so that all might hear.

He then hurled his club high above his head, caught it deftly as it fell. "Who will fight Rorg for his bride? Who will fight Rorg for his kingdom? Speak now, or for another endir, keep silence."

There was a deep grumbling growl behind me, and, turning, I beheld Urg, fangs bared, stepping from his place at the end of the line, whirling his great club. "I will fight Rorg," he shouted in a voice as deep as that of the king-ape.

Rorg appeared surprised—annoyed. For a moment he stood motionless, glowering at his challenger. Then, with a bellow of rage, his club held high in one huge paw and his flint knife gripped in the other, he leaped to the attack.

The club descended in a deadly, whistling arc, but did not connect, for with cat–like quickness Urg leaped to one side and struck back. His club bit deep into Rorg's left shoulder, eliciting a roar of pain and rage from the Rogo, who instantly swung for his legs.

Urg sprang back, but not far enough. The flint–toothed point raked one knee, and blood spurted forth. As he danced about the larger ape, looking for another opening, he limped, and the limp grew more pronounced as the fight progressed.

Again and again Rorg rushed in. How Urg succeeded in evading those rushes, lame as he was, I was unable to understand. Presently his leg became useless, dangling, and he was forced to hop on one foot.

Over the brutal face of Rorg there crept a look of triumph. Deliberately, now, he advanced toward his opponent, forcing him backward until he stood on the very brink of the plateau.

He leaped in, and as Urg swung a slashing blow for his neck, he ducked, at the same time whirling his club in a low, horizontal arc. It caught the challenger halfway between knee and ankle; there was a snap of severed bones, and Urg toppled backward to alight on his head on the rocks seventy–five feet below.

Scarcely had he struck ere the milling horde beneath rushed to the spot, brandishing their flint knives. In less time than it takes to tell, the body had been dismembered, and a snarling group of apes was fighting over the fragments.

Again Rorg leaped in the air, bellowing forth his deep–voiced challenge. Although there were low growls from the ape–warriors standing behind me, none answered the challenge.

"Who will fight Rorg for his bride and his kingdom?" The final challenge was flung out by the victorious king–ape as he looked triumphantly about him. "Speak now, or…"

"I'll fight you, Rorg," I said, drawing club and knife and stepping in front of the giant. As I did so I caught a fleeting glimpse of Taliboz and Loralie. On the face of the traitor was pleased anticipation. The eyes of the princess showed surprise, and something more. Incredible as it appeared from her recent actions, it was undoubtedly concern for my safety.

But these were only fleeting impressions.

Rorg stared incredulously down at me for a moment, evidently unable to believe that I had actually challenged the king of the cave–apes. Then he struck at me quickly, but not exerting his full strength, as if I were some insect annoying him.

Instinctively I used my club as if it had been a sword—parrying the blow with ease and countering with a thrust which bit into his furry abdomen, drawing blood and eliciting a grunt of rage and pain.

Although the club was so constructed that I could not hope to inflict a mortal wound by thrusting the sharp flint teeth with which it was armed, it could and did cause considerable pain and annoyance. As the cave–ape system of fighting was merely that of striking and dodging. I hoped to offset my adversary's enormous advantage of strength and reach by employing the technique of a swordsman.

With an angry bellow, Rorg swung a terrific blow for my legs. Again I parried, and countered with a neck cut which would probably have terminated the engagement in my favor had it not been blocked by one of his huge tusks. The tusk snapped off and clattered to the rock; but as a result, the club wounded him only slightly, adding to his fury.

Foaming at the mouth and gnashing his teeth in his rage, the king–ape beset me with a rain of blows that would have been irresistible to any but a trained swordsman. Splinters and bits of broken flint flew from our clubs as time and again I parried his terrific blows.

After each blow I countered with a cut or thrust, and soon my opponent was bleeding from head to foot; yet his strength and quickness seemed rather to increase with each fresh wound. Had he possessed a swordsman's training, I verily believe that ape would have been invincible on his own planet or any other.

Presently I succeeded in raking him across the forehead with the point of my weapon, so that the blood ran down in his eyes, half blinding him. But he wiped the blood away with the back of one huge paw and countered with a blow, the force of which numbed my wrist and splintered my club into fragments.

I leaped back, then hurled the club handle straight for the great, snarling mouth as he bounded forward to finish me. It struck him in the front teeth, breaking off several and momentarily bewildering him.

In that moment I leaped, and with the fingers of my left hand entwined in the wiry hair of his chest and my legs gripping his waist, I buried my flint knife again and again in his brawny neck. Blood spurted from his pulsing jugular as he endeavored to shake me off, to reach me with his sharp fangs, and to gore me with his single remaining tusk. But his mighty strength was spent—his lifeblood draining.

A quiver shook the giant frame and like some tall tree of the forest felled by the woodman's axe, he toppled backward, crashing to the ground.

Leaping quickly to my feet, I seized the club of the fallen ape–monarch and, brandishing it aloft, said, "Rorg is dead, and Zinlo is king. Who will fight Zinlo? Who will be next to die?"

From the throats of several of the ape–warriors in the semicircle from which Urg had come, came low growls, but none advanced, and the growls subsided as I singled out in turn with my gaze each of the truculent ones who had voiced them.

Far below me, the mob of apes was clamoring, "Meat! We want our meat!"

I knew that, spent as I was, the enormous body of Rorg was more than I could raise aloft and hurl to the mob below, so I had recourse to an old wrestling trick. Seizing the limp right arm of the fallen king–ape, I dragged the body to the edge of the cliff. Then, bringing the arm over my shoulder in an application of the principle of the lever, I heaved the remains of Rorg over my head.

A moment later the milling beasts below were tearing the carcass to pieces, snarling and snapping over their feast. This custom, I afterward learned, had been established in consequence of the belief that the flesh of a strong, brave individual would confer strength and bravery on the one who devoured it.

Again I brandished my club aloft, shouting, "Who will fight Zinlo for his kingdom? Speak now, or keep silence for another endir."

This time I heard not even a single growl from the warriors on the cone top.

An old warrior who had lost both tusks, an ear, and several of his fingers, stepped from the ranks and advanced to the cliff edge. "Rorg is dead," he announced. "Farewell to Rorg."

Following his words, a peculiar, quavering cry went up from the throats of the thousands of apes congregated in the crater, as well as from those on the plateau. So weird and mournful did it sound that I shivered involuntarily.

As the last plaintive notes died away, the old warrior shouted, "Zinlo is king. Hail, Zinlo!"

A deafening din followed as the ape–horde, brandishing knives and clubs aloft and clattering them together, cried, "Hail, Zinlo!"

I turned in triumph toward the spot where Taliboz and Loralie had been seated, intending to assure the princess that it would not be necessary now for her to marry the king of the cave–apes. To my surprise, I saw that both of them had disappeared. The two huge females who had been guarding them sat, side by side, slumped against a large boulder, their chins sunk forward on their hairy chests.

Bounding forward I seized one of the she–apes by the shoulder and shook her, shouting, "Where are your prisoners?"

Her limp body sagged forward, falling on the ground. The second female, when shaken, showed some signs of returning consciousness.

"What happened?" I asked. "Where are your prisoners?"

Weakly she pointed to a needlelike glass sliver embedded in her arm. Extracting it, I instantly recognized it for a tork projectile of the type which temporarily paralyzes its victim. In the arm of the other, a similar projectile was embedded.

Although he had been disarmed by the apes, it was evident that Taliboz had managed to keep his ammunition belt, and that during the excitement of my fight with Rorg, he had found the opportunity to paralyze the two female guards and slip away with the princess.

That she had gone with him willingly I could not doubt, for she had made no outcry, and her previous treatment of me had led me to believe that she would sooner have accepted Rorg for a mate than me.

I turned away, the sweetness of victory grown bitter in my mouth. I was about to enter the runway which led to the cave below, when a small, glittering object attracted my attention. Stooping, I picked it up and examined it minutely for a moment. Then a great light dawned on me.

Chapter VII

HURRYING DOWN the runway into the great cave below, I was about to rush out into the daylight to examine the small object I had found, when a long, muscular arm suddenly went about my shoulders, my head was crushed against a soft, furry breast, and a pair of pendulous lips caressed my cheek.

With the heel of my hand I pushed the face of a she-ape from mine and broke her embrace. Surprised, I recognized Chixa. She advanced toward me again, arms outstretched, but I motioned her off.

"Stand back," I warned her. "What do you mean by this familiarity?"

"But I am your mate," replied Chixa. "You have slain Rorg and the other she has run away. Rorg chose me for his mate before the food-woman came."

"Rorg chose his own mates, and I'll choose mine," I retorted. "What's this you say about the other she running away?"

"The food-man and she came down the runway together. I let them escape. I did not want the food-woman to take my place."

"But how could they escape when the place is surrounded?"

"The food-man knew of the inner passageway," replied Chixa. "I showed him where it was...Am I not as comely as the other shes of my people?"

"No doubt you are the most comely, Chixa, but I will never mate with a cave-ape. You say this she went willingly with the food-man?"

"She did. I think they will be mates."

"Chixa," I said, walking to the entrance and examining the small glittering object that I had picked up, "you have lied to me."

"I lied," admitted Chixa, not one whit abashed, "but how do you know?

You must be a sorcerer, as Graak said."

"I know by this small, broken glass needle, one end of which is stained with blood," I replied. "Call it magic, if you like, but this needle tells me that the she was carried away by the food–man."

"It is even as you say," conceded Chixa. "She was unconscious from the magic of the food–man, and her arm was bleeding."

"Show me the entrance to the inner passageway," I commanded.

Chixa sulked, and crouched in a corner.

"Show me the entrance," I said again, "or I will kill you by magic and feed you to the crowd outside."

Evidently the threat to kill her by magic—the fear of the unknown—was more potent than any ordinary death threat could possibly have been, for she rose, and, walking to the back of the cave, heaved a great slab of rock to one side, disclosing the dark mouth of a runway.

"It was this way they went," she said, "but you will never find them. By this time they will have taken trails where none but our greatest trackers could scent them out.

"Who is your best tracker?"

"Graak is the greatest of them all."

"Go instantly," I commanded, "and bring Graak to me. See that my command is carried out at once, or my magic will follow and slay you."

"I go," she responded fearfully, and hurried from the cave.

I fidgeted impatiently until she returned with Graak, who unhesitatingly offered to obey his new Rogo. Stooping, he entered the passageway. I hurried after him with my hands outstretched in the inky blackness in front of me to prevent dashing myself against the curving walls. We must have gone two miles in this manner before twilight loomed ahead, followed by daylight, and we emerged in the open air on a narrow shelf of rock against which the topmost fronds of a giant tree fern brushed. Around and beyond this mighty fern stretched a forest of its fellows, coming up to the very edge of the mountains that held the homes of the cave-apes.

Graak sniffed the air for a moment, then leaped for the nearest fern frond, which sagged beneath his weight as he caught it with both hands. His great body swung precariously a full seventy feet above the ground as he went up the slanting frond, hand over hand, until he reached the trunk. After sniffing at this for a moment, he descended, feet first, to the ground.

I followed his example, making much more work of it than he, and descending so slowly that he stamped impatiently before I reached the ground. I wondered how Taliboz had been able to negotiate this route with his inert burden until I noticed a long, slender cord dangling from the end of one of the fern fronds, its lower end about ten feet from the ground. The traitorous noble had evidently lowered Loralie by means of this cord to within reach of the ground, where he had evidently cut her loose and carried her off.

While Graak fidgeted impatiently, I leaped and caught the end of the cord. I called him to help me, and together we pulled until the frond broke off and came crashing to the ground. With my flint knife I quickly cut the cord from the branch and, coiling it about my body, told Graak to proceed. Feeling that we might have a journey ahead of us, I thought of several ways in which the cord might be useful.

We had not gone more than a mile in the fern forest when the cave-ape pointed to a set of smaller footprints beside Taliboz's and said, "The she walked from here."

Recovering at this point from the paralysis induced by the tork projectile, she had gone on with her abductor, willingly or not.

Although the footprints led at first toward the west, they presently began to turn southwest, toward the coast.

For many hours we followed the trail without food or drink; then Graak stopped in a clump of bush ferns which furnished us pure, fresh water. He next plucked some sporepods, cracking them with his teeth. I split some open with my knife. They had a pleasant, nutlike flavor.

We resumed our journey until the advent of sudden darkness, when we climbed into the leaf-crown of a tall tree fern to pass the night there.

Graak fell asleep at once, but I could not. No sooner had darkness descended on the forest than the night–roaming carnivora were astir, making the night hideous with their cries—howling awoos, the horrid, mirthless laughter of hyenalike hahoes, the terrific roars of marmelots, the death–cries of the victims.

I think the gentle rocking of the trees, together with the rustling of the countless millions of fern leaves, lulled me to slumber. At any rate, I was awakened by the great hairy paw of Graak pulling at my arm, which I had thrown across my face—a habit of mine while sleeping. "The light has come," he said, "and Graak is hungry. Let us find food and be gone."

As I followed him down the rough, scaly trunk, I was struck by the contrast of the daylight sounds. I could hear only the buzzing of insects, the silvery toned warbling of the awakened songbirds, the occasional snort or grunt of some herbivore feeding, and the peculiar squawking cries of the queer bird–reptiles called aurks.

Graak and I had only traveled a short distance on the trail when he suddenly stiffened and, looking upward, said, "Good food! A ptang!"

Following the direction of his gaze, I saw a large, hairless slothlike creature hanging upside down on a thick fern frond which bent downward beneath its weight. The ptang was unconcernedly munching leaves without so much as a glance in our direction.

The cave–ape bounded to the base of the tree and quickly ascended, to climb out on the limb where the stupid creature was feeding, paying no attention to the approaching danger.

Graak swung by a prehensile foot and hand, and struck with his saw–edged club, laying the side of the creature's head wide open at the first blow. It ceased its feeding, but did not attempt either to fight or run away, though its powerful legs were armed with long, hooked claws. Again Graak swung his club. The animal's head hung limply downward and a shiver ran through its frame.

Replacing his club in his belt string, the cave–ape drew his flint knife and pried the hooked claws one by one from their grip on the limb. The ptang crashed downward through the branches to the ground.

When we had eaten our fill, the ape and I each cut off as large a portion of the animal as could conveniently be carried, and started once more on the trail.

We had not gone far when Graak pointed out a place where Taliboz and the princess had stopped to eat, the night before. A little farther on the trail, we came to the base of a large tree fern in whose leaf crown they had passed the night. Evidently they were not more than an hour ahead of us.

As we hurried forward and the scent grew stronger and stronger, the cave-ape showed all the excitement of a hound on a fresh game trail—which it was, to his mind.

Presently he stopped, tensely alert, sniffing and listening.

"What is it?" I asked in a whisper.

"A marmelot follows them," replied Graak, pointing to the footprints in the leaf mold.

Looking down, I saw, sometimes between their tracks, sometimes obliterating part of them, the spoor of a gigantic feline, so heavy that it sank to a depth of nearly a foot with each step.

Then carne the scream of a woman in deadly terror, only a short distance ahead, followed by the crashing of underbrush and a terrific rumbling growl which I recognized only too well.

Graak instantly took to the trees, but I unlimbered my club and knife and dashed forward.

Hurrying as fast as I could in the soft leaf mold, dodging through fern-brakes and tripping over creepers, I presently floundered out into a little glade where a most fearsome sight met my eyes.

Rolling about on the ground, snapping, tearing and clawing at everything that came within its reach, was a magnificent marmelot, apparently in its death throes.

I had not taken three steps before the creature quivered, subsided, and lay still.

Looking about for the princess and her abductor, I was startled by a warning cry from almost directly above me, "Zinlo! Behind you!" It was the voice of Loralie.

Whirling, I saw Taliboz standing behind the broad trunk of a tree fern. In his left hand he held an object which I recognized as a clip for tork projectiles. Balanced in his right hand with its base against his palm and its length parallel with his fingers was one of the needle–like glass bullets, ready to throw. Even as I looked, he hurled it straight for my face.

I ducked my head just in time, heard the bullet strike a fern trunk behind me, and sprang forward. But he quickly pulled another from the clip and I saw that I could not reach him in time to use my weapons; nor could I, close as I was, again hope to avoid the throw by dodging.

With a grin of triumph on his features, he swung back his arm, poised it for a moment to get his aim, then brought it swiftly forward, his fingers pointing directly at my breast.

"Die, stripling!" he grated between clenched teeth.

But a strange thing happened. Instead of feeling the sting of the needle in my breast, I saw him go limp and slump down in his tracks.

I learned the cause as I bent over to, examine him. The needle bullet which he had intended for my breast had pierced one of his fingers instead. Rolling him over, I took his tork ammunition belt and buckled it about my own waist. I picked up the clip which he had dropped when he fell, and, closing the ejector, replaced it in the belt.

Then I looked up in the direction from which the warning voice of Loralie had come down to me. For a moment only I saw her beautiful face peering down at me between the parted fronds of a leaf–crown. Then a huge hairy arm reached downward, encircled her slender waist, and drew her backward. She cried out in deadly fear as the parted fronds snapped back in place, hiding her from view.

I caught a glimpse of Graak mounting one of the rope–like vines; beneath his left arm he carried the drooping form of Loralie. Then they both disappeared into the thick tangle of vegetation above.

"Stop, Graak!" I called. "Come back, or I will slay you with my magic."

No answer.

I leaped for the nearest fern trunk, intent on following, when suddenly, without the slightest hint of warning, a long sinuous object whipped through the air and coiled itself about me. With its deadly fangs gleaming in gaping jaws quite close to my face, and cloven tongue darting forth menacingly, the glistening beady eyes of a gigantic whistling serpent stared hypnotically into mine.

Swiftly, relentlessly, the mighty coils tightened about my body while the horrible head moved rhythmically back and forth, just above my face. My club was caught beneath the scaly folds of my assailant, but I managed to jerk my flint knife free, and with this I struck at the swaying, silver–white throat. But the covering was tougher than I had thought, and I only succeeded in chipping off a few scales.

The muscular coils that encircled me grew tighter. It seemed to me that my ribs must crack at any moment. My breathing was reduced to short, spasmodic gasps.

Then I thought of the tork projectiles. With my flint knife I pried the ammunition belt up from beneath an encircling coil. Quickly extracting a clip, I opened the ejector, pressed the button, and a small, sharp needle popped out. I slid it under the edge of a scale and pressed. Scarcely had I done so when the crushing folds about me began to relax; the swaying head dropped limply downward, and I tugged and wriggled until I was free.

Still gasping for breath, I closed the safety catch of the clip and replaced it in my belt. I noticed that it was marked in patoa: "Tork projectiles—deadly."

As soon as I was able to breathe with reasonable normality once more, I climbed the tallest tree fern in the vicinity, and from its lofty leaf–crown looked out over the tree–tops in the hope of locating Graak and the princess. But although I scanned the forest in every direction I could not catch sight of them.

Far back toward the northeast, the mountains of the cave–apes were barely discernible through the gray–blue mistiness that hung over the jungle. Toward the southwest, and closer, was another mountain range—gray, forbidding peaks much higher than those of the cave–apes.

As he was, by nature, a cave dweller, I decided that Graak would eventually seek a mountain home. Having disobeyed me, King of the Cave–Apes, he would not dare return to the mountains of his tribe. I might very logically expect him to head for the other mountains. When I had caught my last glimpse of him he actually was starting toward the southwest. I decided to travel that way, zigzagging across my plotted course in the hope that I might eventually pick up his trail.

Having made my decision, I descended to the ground and set out toward the unknown mountains.

I was in the middle of my second zigzag toward the south when I came across the trail of Graak. Dainty but significant beside those of the cave–ape were the tiny footprints of Loralie. As I followed the trail I twice saw the records of her attempts at escape—where she had tried to run away, but had been caught.

Now travel became far more difficult. My first warning of the changed terrain was when I sank hip–deep into a sticky quagmire, only saving myself from complete immersion in the soft mud by grasping a stout vine that hung across my path, and swinging up into firmer ground. I noticed that fungi and lichens were beginning to predominate.

Gradually the tree ferns and cycads were replaced by gigantic toadstools of variegated forms and colors, and huge morels, some of which reared their cone–like heads more than fifty feet in the air. Jointed reeds rattled like skeletons in the breeze; lichens upholstered rotted stumps and fallen logs, and algae filled the treacherous, stagnant pools that grew more numerous as I advanced, making it difficult to tell which was the water and which the land.

It was comforting for me to know that the flight of Graak was being even more retarded than mine. He had to test each bit of ground before treading on it, while I had but to follow his footsteps.

Suddenly I heard, only a short distance ahead of me, the angry roar of the cave–ape, followed by a woman's scream of terror.

At first I thought Graak had sighted me, and I dashed forward to meet him with club and knife ready. But before I had taken a dozen steps I heard his voice raised in a howl of pain, and soon he was alternately bellowing and snarling as if in intense agony.

I caught sight of Graak and the princess at the same time. The ape, his fierce cries now reduced to mere whimpering, was on his back surrounded by a half dozen of the strangest and most horrifying creatures I have ever seen.

Writhing, squirming, extending, contracting, they had no set form, but could change themselves instantly from limbless, egg–shaped bodies three feet long to the semblance of snakes fifteen feet in length, or any of the intermediate lengths between the two. They were clinging to the fallen cave–ape with grotesque, three–cornered sucker mouths, and from the edges of some of them I could see blood dripping.

Before I could reach him, Graak's whimpering subsided, his struggles ceased, and I knew that he was beyond help. His assailants, finding him quiescent, settled down uniformly in the shape of extended ovoids about four feet in length, to drain the rest of his blood.

From a position of temporary safety, the princess looked down in horror. She was on the umbrella–like top of a toadstool about fifteen feet in height, evidently having been tossed there by Graak when he had been attacked, for there was no way she could have reached that point unassisted. Climbing rapidly toward her were two more of the hideous things, leaving slimy trails on the stem.

Bounding forward, I swung my club at the nearest creature, expecting to cut it in two with the sharp, saw–edge of my weapon. To my surprise and consternation, the club failed to make the slightest impression, but bounced off as if it had struck extremely springy rubber, and nearly flew from my grasp.

The hideous head with its three–cornered suck mouth was instantly extended toward me, and again I struck—this time from the side. Although the blow made no more impression on the tough skin of the creature than before, it broke the hold of the thing on the stem of the mushroom and sent it whirling and writhing a full twenty feet away.

The other thing on the stem stretched out to seize me, but I dealt it a backhand blow which sent it squirming and wriggling in the opposite direction.

A quick glance around showed me that the surrounding marsh was literally alive with these horrible creatures. Evidently excited by the sound of the conflict—or possibly by the smell of blood—they erected ugly swaying heads to investigate, then came crawling toward us, leaving slimy trails in their wake.

There was only one thing for me to do in order to save the princess, or even to save myself: I must find a way to get to the top of the toadstool with her. But this was a good fifteen feet from the ground, and the marshy soil was not particularly conducive to high jumping, as it clung to the feet with each step.

As I looked about for some means wherewith to accomplish my purpose the ring of attackers closed in on me. Then came an inspiration. About twenty feet from the toadstool on which the girl stood was a clump of huge, jointed reedlike growths. Several of them, which reached to a height of more than forty feet, bent slightly toward it.

I managed to reach them just ahead of the advancing army of attackers and climbed the largest one with an agility of which I had never even imagined myself capable. One of the slimy things that sought my lifeblood instantly wound its body around the reed and followed, then another and another, until the stalk below me was covered with their snaky forms.

As I climbed upward, the reed gradually bent over toward the top of the toadstool, so that when I reached a height of a little over thirty feet, I was directly above it. Swinging my legs free, I hung on for a moment with my hands, then let go. As I alighted on the center of the toadstool cap, the reed shot upward like a steel spring, hurling its slimy occupants far out across the marsh as if they had been shot from a catapult.

No sooner had I alighted than there was a cry of terror from Princess Loralie. Turning, I saw her crouching in fear beneath the ugly head of one of our attackers, its neck arched and its three–cornered sucking mouth, armed with thousands of razor–sharp cutting teeth, ready to strike.

I swung my club, knocking the thing to the ground, but no sooner had I done so than another came up over the edge of the toadstool, quickly followed by two more. Soon the entire rim became alive with the swaying, wriggling heads, and I was kept busy every second of the time knocking them back to the ground.

"Give me your club, Prince Zinlo," said Loralie after I had been at this strenuous work for some time, "and let me help you. If we take turns with rests between for each, we can last longer. The swamp dwellers are persistent, and we are doomed, it seems—but let us fight while life lasts."

"I am not tired," I insisted, rather breathlessly, but she came and seized the club, making it necessary for me either to use force with her or surrender it. I yielded, watching her to see if she could manage it. Despite her small size she proved surprisingly strong.

But she soon grew weary, and I took the club once more. It was a hopeless fight; day was fast waning, and in the black, moonless darkness of Venus we would soon be dragged down to meet the fate of the bloodless carcass that had once been Graak, now staring sightlessly up into the leaden sky.

Chapter VIII

I WAS running around the rim of the toadstool cap, knocking off the slimy things that sought to drink our blood, and Princess Loralie was crouching fearfully in the center, when suddenly I heard a crashing and splashing through the marsh behind me, accompanied by queer noises that sounded much like a combination of a bleat and a bellow.

Glancing back for a moment between gasps, I saw coming toward us an immense humpbacked reptile sinking flankdeep in the watery ooze with each step as it crashed through the reeds in its apparent endeavor to escape from some mortal enemy, and uttering the queer cries of distress I had heard. I could see its long snakelike neck curved back as, with its small jaws it would jerk the swamp creatures first from one side then the other.

Coincident with the appearance of this huge reptile, the heads of the swamp dwellers stopped reappearing above the edge of our toadstool cap. They had abandoned their attack on us in favor of the larger quarry.

Thicker and thicker they swarmed around the great dinosaur. For every blood–hungry thing the giant lizard tossed in the air, at least ten squirmed up to fasten their sucker mouths on its heaving sides, until the reptile's back resembled the wave–tossed bottom of a capsized ship covered with immense barnacles.

Gradually the speed of the great beast slowed down. It stopped. Then its struggles grew weaker, and the doomed saurian uttered a final cry and sank down in the ooze.

So absorbed had I been in this titanic battle that I had momentarily forgotten our own danger.

"Our enemies have momentarily forgotten us," I said then. "Shall we make a dash for liberty?"

"It is our only chance," she replied.

Swinging over the edge of the toadstool, I dropped to the ground. Loralie swung her small, athletic body over the edge as I had done, and dropped into my waiting arms.

As I stood there, ankle deep in the ooze with that shapely young form close to me, I suddenly forgot our danger—forgot everything except that she lay there in my arms, her head thrown back, glorious dark eyes that were pools of lambent flame looking up into mine. But when, intoxicated with her nearness, I would have crushed her to me, she suddenly twisted free from my arms and ran, leaping lightly as a startled fawn in the direction of the mountains to the southwest.

Club in hand I followed her as closely as I could, meanwhile keeping a sharp lookout for swamp dwellers. But they were too busy feasting.

As we approached the foothills the ground became drier and firmer, and the character of the vegetation once more underwent a gradual change; cycads and low-growing conifers were mostly in evidence. Soon we were climbing steep hillsides, with the ground continually becoming more rugged and the vegetation more sparse.

During our progress Loralie had not addressed a word to me, or noticed my presence in any way. I felt I must have offended her by holding her over-long in my arms. Yet for that fleeting moment I would have sworn I had seen in her starry eyes the reflection of emotions akin to my own, and quite unlike her unnatural aversion to me in the caves of the ape.

When we arrived in a small isolated copse of water ferns, I decided it was time to halt for rest and refreshment.

"Here are food and drink," I said. "Let us stop for a while."

Without answering, she sank down wearily on a mound of soft moss and turning, buried her face in her arms. In a moment she began weeping softly.

I broke off a branch of water fern and knelt beside her, trying to get her to sit up.

"Don't touch me!" she wailed. "Go away."

"Oh, very well," I snapped, and ate and drank by myself—without much appetite. Then, I set about equipping myself with more efficient weapons.

I soon fashioned a bow, which I strung with a piece of the tough cord I had brought with me. Some reeds which I had gathered en route I made into arrows by tipping them with slivers of stone bound in place with the cord. I bound bits of fern leaf in place of feathers. A quiver I made from ptang–hide which was wrapped around the piece of meat I had brought with me.

Several hours elapsed in these pursuits, and my too temperamental companion had in the interval sobbed herself to sleep.

I had scarcely finished cooking some ptang meat when I saw the princess stir and open her eyes. For a moment she seemed startled by the strangeness of her surroundings. Then she sat up, and catching the appetizing scent of my roasting meat, looked hungrily toward it—then resolutely away.

"The Prince of Olba," I said, "would be greatly honored if the Princess of Tyrhana would join him at dinner. The royal butler is about to serve."

Despite her attempt at severity, I saw a slight smile play around the corners of her adorable little mouth. Then she turned her head away once more.

Placing my roast on some broad, clean leaves which I had spread over the moss for the purpose, I walked over to where she sat.

"I say, young lady," I remarked severely. "Don't you think you have carried this foolish perversity of yours about far enough? I can't imagine what makes you act like a badly spoiled child. I've a notion to spank you."

She tried to maintain her dignity, but I saw her lips quivering.

"Forgive me," I said. "Perhaps it is I who am wrong. If I have done anything to hurt your feelings, I'm sincerely sorry. I am not desirous of forcing my attentions on you, but I can't leave you alone in this wilderness. You make it hard, extremely hard for me to be of service to you."

She looked up at me, her beautiful eyes brimming—tears clinging to the long dark lashes. "You are so kind, and so brave. I wish those other things were not true."

"What other things?" I asked in surprise, sitting down beside her. "Has someone been talking about me?"

"I cannot betray those who have reposed confidence in me," she said, "nor can I doubt the testimony of many witnesses. Yet it does not seem possible."

"I'm sure I don't understand what you are driving at. Pray tell me of what monstrous crime I am accused, and permit me at least a chance to defend my character."

"You were accused...Oh, I cannot say it!" She looked at me reproachfully, then turned her head away and swallowed bard to keep from crying.

"It must have been horrible. Won't you tell me what it was?"

"Of making love to that Chixa," she faltered.

The evidence might seem to point that way, I realized, particularly if it were distorted by someone bent on maligning my character. I quickly told her how I had won the she–ape's weapons and my freedom. "Do you not believe me?" I demanded at last.

"On this matter I believe you," she answered with some relief, "but there is still that other affair."

"What other affair?" I asked.

"Your affair with the young sister of Taliboz. Why did you betray that trusting child—betray her and run away—so that her brother must needs come after you to bring you back at the point of a tork? It was dastardly—cowardly. I denied it—fought against believing it, but there were so many witnesses I was at last convinced."

"If Taliboz has a sister, I do not know it, nor have I ever seen her. This story was fabricated from whole cloth. There is not even seeming evidence in this case as there was with Chixa."

"But Taliboz himself told me," she insisted, "and five of his men substantiated his story at various times. I wanted to disbelieve this thing, but what could I do?"

"You were convinced of a monstrous falsehood, for which Taliboz will one day answer, as he will for his numerous other crimes—if he has not already answered, back there in the fern forest, to some jungle creature. I swear to you that if Taliboz has a sister I do not even know of her existence."

"It seems strange," she answered, "that the sister of an illustrious noble of Olba should be unknown to the Crown Prince. Surely she must have been much at court."

"Perhaps she was. Never having been there myself, I cannot say."

She looked at me in amazement—unbelief so clearly written on her features that I saw that I had gone too far. I must either tell all now, or have nothing believed.

"In order that you may understand this singular statement," I said, "I am going to tell you who I really am."

"No doubt you are a reincarnation of the god Thorth. Pray do not weary me further with lies."

"The story I am going to tell," I answered, "will tax your credulity to the uttermost, yet I hope some day to be able to prove it to you. I am not of Olba, nor even of this planet."

I explained to her, as best I could, how I had been transported from Mars to Earth and thence to Venus–Zarovia. To my surprise, she seemed not only credulous, but actually well versed on the subject.

"You seem to know more about these phenomena than most scientists," I said.

"There is a reason for my intense interest in the subject," she replied. "My uncle Bovard is one of the greatest scientists on all Zarovia. There is but one who is considered greater than he."

"Vorn Vangal?"

"Yes, but how did you know?"

"Vorn Vangal," I answered, "is Dr. Morgan's Zarovian ally, the man who made it possible for me to come to this planet."

"Dr. Morgan? What an uncivilized sound the name has! Vorn Vangal I know well."

"Then you believe my story?" I asked.

"Implicitly." And she smiled thrillingly at me.

"And you know Taliboz was lying?"

"Of course. Are you going to sit there and question me all day, or will you have the royal butler serve dinner? I am famished."

The roast had grown cold but was nonetheless delicious. I carved as best I could with my flint knife, and we made out very well, finishing up with the contents of a few spore pods, washed down with drafts of cold water from the fronds of the water fern.

"And now," I said, when we had finished dinner, "we must look about for a place of shelter from the night-moving meat-eaters."

There were many caverns in the rocky hillsides, but the mouths were too large or too numerous to be barricaded. And an unbarricaded cave in the Zarovian wilderness would prove to be a trap.

We traveled far before we found a cave that seemed suited to our purpose. Without taking time to explore its interior—for I knew that the sudden darkness would soon be upon us—made haste to collect heavy rocks for the doorway, delegating Loralie, meanwhile, to gather sticks for fuel which I intended to keep in the cave as a fiery defense against possible attackers.

Darkness caught us with our labors unfinished, and I kindled a small fire just outside the cave mouth that we might complete our work by its light.

I was just rolling up the great stone which was to finish my barricade when the hideous roar of a marmelot sounded near by. It was taken up, a moment later, by others of its kind, until the echoing hills resounded with the thunderous cries of these fierce beasts.

"Quick!" I called to Loralie. "Into the cave with you!"

She started in, then backed out in terror. "There's something in there now, and it's coming out after us."

Then, as the frightened girl cowered against me, I heard a hoarse, booming croak from the cave and saw two glowing, menacing eyes moving toward us from the darkness of the interior.

Chapter IX

STANDING WITHIN the ring of light cast by our small fire, with Loralie crouching fearfully at my feet, I fitted an arrow to my bowstring. I drew it back to the head, took careful aim between the two glowing eyes that were advancing from the dark interior of the cave, and let fly.

Immediately after the twang of the bow there came a deep bellow of rage from the interior of the cave.

As I fitted a second arrow in place, there was a terrific roar behind me. Turning, I beheld the gleaming eyes of a marmelot not more than fifty feet distant. I let fly, and the arrow struck the huge feline just as the enraged cave creature came forth.

Prepared as I was for the appearance of one of the fierce creatures of the Zarovian jungle, a chill of horror ran down my spine when the grotesque tenant of the cave waddled out into the light.

It was obviously a reptile—not an animal as I had supposed. Although its entire length was not more than six feet, fully two-thirds of that length was mouth—enormous jaws four feet long and a yard across, armed with row upon row of sharp, back-curved teeth. The other third was a round sack, or pouch, attached to the back of the head.

It walked on two short, thick legs growing from beneath its ears, each armed with three sharp talons. There were no forelegs. Both head and body bristled with a profusion of sharp spines like those of a horned toad.

"A kroger!" cried Loralie. "We are lost!"

As the thing charged toward us with enormous jaws distended, I heard the marmelot bounding through the brush from the opposite direction.

"Come," I cried, taking the girl's hand. Together we leaped across the fire and into the shadow of the bushes beyond. Scarcely had we gained this place of temporary safety ere the two formidable creatures, beast and reptile, met on the spot where we had been standing.

The marmelot, apparently surprised at being confronted by this strange anomaly, stopped, spat, and arched its back like a startled cat. But the kroger, undaunted at sight of the huge king of the jungles, which was easily twice its size, charged on. With a snap of its immense jaws, the reptile took in at one bite the head and neck of the mighty carnivore.

Like a cat caught in a salmon tin, the marmelot alternately shook its head, clawed at the scaly throat, or belly—I know not which to call it—and ran blindly about. Presently it rolled over on its back, and drawing the round body of the kroger toward it with its two front legs, literally scratched it to ribbons with its terrible hind claws. Yet the immense jaws held firmly, inexorably; in fact, they seemed to be clamping down tighter and tighter all the time, sinking more deeply into the flesh of the great feline with every move it made.

The struggles of the combatants presently grew weaker, but as the sounds of battle died down the noises in the fern brakes around us grew closer and more alarming. Evidently attracted by the sounds of battle or the smell of blood, the denizens of the hills drew nearer and nearer in an evernarrowing circle. The weird howling of the awoos, mingled with the grisly laughter of the hahoes and the cries of other night–roving beasts, produced a most uncanny effect.

If we did not find shelter soon, our bodies would go to appease the flesh–hunger of one or another of these hunters.

Warning Loralie to keep out of sight in the bushes, I dashed over to the fire, seized a burning brand and hurled it into the cave. As nothing charged out after me, I peered in. By the flickering light of the burning stick I could see that the cave was small and apparently empty, except for a pile of dry fern fronds against the back wall.

Entering, I picked up the torch and investigated this. It proved to be a nest about four feet across, in the center of which was a round egg, covered with a membranous shell mottled green and yellow—the same color as the outer scales of the kroger.

Flurrying out of the cave once more, I softly called to my companion. "Carry the fuel into the cave at once, while I build our barricade."

While we both worked in frenzied haste, the sounds in the surrounding darkness grew ominously closer. The struggles of the marmelot and kroger had ceased altogether, and our fire was burning low.

Perspiring from every pore with my strenuous labor, I presently got the cave mouth closed except for a narrow hole on one side barely large enough to admit the body of a man.

Loralie had meanwhile carried all of the fuel into the cave and was waiting for me in its dark interior.

Seizing a flaming faggot from the remains of the fire, I squeezed through the narrow opening, then lifted into place the rock I had reserved for the purpose while the princess held the torch for me. Scarcely had I done this ere a half dozen lean gray forms bounded into the glow that was shed by the last few coals of our fire and began tearing at the two mighty carcasses which were locked in a death embrace beside it.

As I watched through the interstices between the rocks, I saw that these were awoos. The more cowardly hahoes soon joined them, and there ensued a fierce medley of growling, snapping and snarling as the beasts fought over their bloody feast.

Because there was no way of ventilating our cave, I disliked building a fire inside; but I felt constrained to do so when a huge hahoe came sniffing up to our rock barrier, then threw back its head and gave vent to the horrid cry which gives it its name. I piled a few faggots against the barricade and lighted them with the flaming brand I still held. It was well I did so, for the cry of the first brute quickly brought a half dozen others and they began sniffing and scratching at the loosely piled rocks.

The smoke nearly strangled us at first, and got in our eyes, making tears stream down our cheeks. But as it billowed out between the crevices in the barrier the besieging beasts sneezed and backed away.

When the moisture had burned out of the fuel it smoked less, and I found that by feeding the fire gradually I could cut its smoking down to a minimum which, though still disagreeable, was bearable.

Glancing across the fire at my companion, I was about to speak to her when I saw that, in spite of her fear, exhaustion had claimed her, and she slept. She lay on her side, her tousled head pillowed on one white arm, her seductive curves outlined in the flickering firelight against the smoky background of the cave's interior.

Despite the tremendous din outside the cave, I presently felt myself growing drowsy. Twice I caught myself wearily nodding, only being able to rouse with an effort at thought of what might happen if our watch fire should go out.

Taking a three–foot length of fern frond, I thrust one end into the fire and laid my hand over the other. At the rate these fronds burned I should catch ten minutes or more of sleep before the flames should reach my hand and awaken me.

I awakened with a start. Daylight was streaming through the crevices in our rock barrier. The fire had ceased to smolder, and the frond on which I had counted to awaken me had gone out more than a foot from my hand. Loralie was still sleeping quietly across from me.

Near the dead embers of our outdoor fire lay the bones of the marmelot and the kroger, picked clean. The vegetation was torn, trampled and spotted with blood, but of the flesheaters that had threatened us the night before I saw no other sign.

Only a short distance away I saw a large clump of water ferns, and toward this I made my way in quest of food and drink. I found these useful shrubs heavily laden with spore pods and, after a refreshing drink, pulled up a number of fronds to take back with me.

As I was walking back toward the cave I caught sight of a small animal browsing on the steep hillside above me. Silently putting down my water–filled fronds, I extracted bow and arrow from my quiver, took careful aim at the animal, and loosed a shaft. Struck just behind the shoulder and pierced clear through, it gave a piteous bleat, sank to its knees, then rolled over and came tumbling down the hillside to fall dead at my feet.

It was a wild frella, one of the hairless, sheeplike creatures which are such highly prized food animals on Venus. I had already tasted the flesh of the domestic variety in the Black Tower.

After returning to the cave mouth with the spoils of my brief excursion, I kindled a new fire on the dead embers of the old one outside, and soon the appetizing aroma of grilling frella steak filled the morning air.

Stepping into the semidarkness of the interior I saw that Loralie was already awake and intently watching the large nest in the rear. "I heard something move back there," she whispered, "and I'm afraid."

Club in hand, I advanced toward the nest. As I did so I heard a peculiar scratching sound which seemed to come from the center where the round egg lay. Yet I could detect no sign of any movement.

Reassured by my presence, the princess came up beside me and peered into the nest. "What can it be?"

Before I could reply, her question was answered from the nest itself. The egg split open and a tiny kroger—like the one slain by the marmelot in every respect except size—rolled out, got unsteadily to its feet, and blinked inquiringly up at us, cocking its head to one side.

I swung my club aloft, bent on quickly dispatching this miniature monstrosity, but the princess caught my arm. "Don't you dare hurt that poor little thing."

The kroger toddled toward her, balanced itself on the edge of the nest, and uttered a rasping, mournful croak.

"The darling!" exclaimed Loralie. "I believe it likes me. Isn't it cute?"

"Cute! It's hideous. I could choke it—if it had a throat"

"Brute! How could you do such a thing?"

"I'm brute enough to be thoroughly hungry," I answered, "and the royal butler is about to serve breakfast. Will you join me or stay here and play with this walking nightmare?"

She held out her hand to the kroger, which instantly opened its enormous mouth to full capacity, and gave vent to a series of high-pitched croaks. "Poor little orphan, it's hungry. I couldn't think of eating a morsel without feeding it. Help it to get down, won't you?"

I extended the flat of my club, intending to shove it beneath the creature's belly, or throat, whichever it might be, and lift it down to the floor. But it sidled away from the weapon—then hopped down by itself and toddled toward the princess. With a little scream of alarm she turned and darted out of the cave, the kroger waddling after her.

I squeezed through the opening as quickly as I could, getting out just in time to see her snatch one of the deliciously grilled frella steaks which I had prepared and toss it into the cavernous maw of the young reptile. It instantly clamped its jaws shut, and dropping the grayish film of its eyelids, settled down beside the princess with its chin between its feet to sleep.

"I told you the little thing was hungry," she said as we sat down to breakfast.

When we had eaten, Loralie insisted that I make her a bow, arrows and quiver. After I had cut a number of reeds into the correct length for arrows I set her to feathering the shafts with bits of fern leaf while I manufactured a number of crude sharp flint slivers for the heads.

After I had a sufficient quantity of these rough tips made, I showed her how to bind them to the shafts, while I scraped, dried, and rubbed with hot fat a section of frella hide for the quiver. While it was hanging by the fire I made a bow.

This work occupied several hours, during which time the kroger slept contentedly beside the princess. When everything was completed and we were ready to resume our journey, the hideous baby reptile promptly woke up and followed us.

As we did not care to run the risk of another attack by the slimy swamp dwellers we planned to follow the mountain range which gradually curved toward the southeast, thus avoiding the marsh and eventually coming out on the coast of the Ropok Ocean. Here we might meet the rescue party of Prince Gadrimel, or failing in this, could try to follow the coast northward to Adonijar.

After about five hours of travel, during which time the princess had been practicing with her new weapons and keeping me busy retrieving arrows, we decided to stop in a small clump of water ferns for food and rest. I had just unslung the haunch of frella meat which I carried and hung it on a fern frond so the young kroger couldn't get it, preparatory to building a fire, when I heard a terrific roar come from over the brow of the hill, followed by the shouting of men, the crashing of underbrush, and intermittent snarls and growls.

I hurried to the hilltop to investigate, the princess running after me and the kroger waddling behind her as fast as its short legs would carry it.

Taking cover behind the bushy fronds of a cycad, I peered down at the scene of strife below. A party of men, about fifty in number, was engaged in a battle with an enormous ramph. The huge, hairless, bear–like creature reared up on its hind feet from time to time, towering above the men around it like a giant among pygmies.

Half a dozen of the men already lay motionless on the ground, yet the others, swarming about the fierce beast, seemed absolutely fearless. They were armed with knives and long, straight–bladed, two–edged swords, and were naked except for their sword–belts, which appeared to be of metal links, and their gleaming, conical helmets or casques.

They were a white–skinned race—too white, I thought, as if they spent nearly all their time indoors. And they wore no beards—an unusual thing on Zarovia, where a beard, cut off square below the chin, was a mark of fashionable manhood.

As I watched, a man darted in to deliver a thrust with his sword. Before he could do so the ramph whipped out with a huge paw and stretched him, crushed and still, on the ground a full twenty feet away. Another man who succeeded in pricking the creature beneath the right shoulder met a like fate.

Instinctively I reached for bow and arrow, but remembered that at that range an arrow could not possibly do more than add to the fury of the beast. Then a scheme came to my mind which I instantly put into execution. Removing an ammunition clip marked Tork Projectiles, Deadly, from the belt I had taken from Taliboz, I extracted one of the needlelike missiles and with a bit of cord, bound it to the head of my arrow.

After replacing the clip in my belt, I took careful aim and released the shaft. It struck the ramph in the shoulder and the deadly virus acted almost instantly; in a few seconds it keeled over, to fight no more.

Apparently mystified at what had killed the great beast, the men clustered curiously about the fallen brute, examining it intently. One pulled the arrow from its shoulder and was instantly surrounded by a group of his comrades, all eager to see and handle it.

"Shall we make ourselves known to them?" I asked the princess, who was peering over my shoulder.

"As you will," she replied. "They seem to be soldiers of a civilized nation, but one I do not recognize. No doubt they will be glad to help us when they know who we are."

I stepped from behind the cycad and shouted the universal Zarovian word for peace—"Dua!"

The entire armed band whirled toward me, and I was horrified at the unhuman quality of their gaze—as if they were more, or less, than men.

Chapter X

THE LEADER of the hunters called out "Dua" and Princess Loralie stepped from her hiding place to my side. Together we walked toward them.

"I am Pangar," said their leader, according us the royal salute in deference to the scarlet we wore. He himself, although not clothed, had a purple band on his metallic helmet and touches of purple on his accouterments which marked him as a member of the nobility.

"I am Zinlo of Olba," I replied, acknowledging his salute, "and this is the Torrogina Loralie of Tyrhana."

"In the name of my royal master, Tandor of Doravia, I bid Your Highness welcome," he said. "Will you accompany me to the palace and permit my emperor the pleasure of greeting you in person?"

"We'll be delighted."

"Your indulgence for a moment, then, while I see if any of my men can be salvaged."

"Salvaged!" I was struck by the peculiarity of the term when applied to men. It brought home to me that there was something extremely odd about these people. The motions of many of them seemed to be quite stiff and awkward—mechanical, that was it—like the motions of marionettes.

Their armor—accouterments and weapons, too—were not made of ordinary metal, as I had first thought, but were constructed from a material which greatly resembled glass. The blades of the swords and daggers were quite transparent. The hilts resembled colored glass.

The helmets were also transparent, except for the colored band at the base of each denoting the status of the wearer. The chain belts and shoulder straps were of the same material, but lined with ramph leather, evidently to prevent their contact with the body.

Pangar bent over one of the fallen men. "Think you can make it?" he asked.

The stricken one spoke weakly. "Power unit is low. Was shorted for a time, but I have it back in place now. If someone can spare some power..."

"Who can spare power?" asked Pangar.

A man stepped up. "I can spare five xads."

"Good." From a hook on his belt, Pangar took two coiled tubes that resembled insulated wires with metal sockets at each end. He inserted an end of each wire in each ear of the fallen man and handed the other two ends to the man standing. The latter instantly inserted an end in each ear, meanwhile watching an indicator which was strapped to his wrist. Presently he jerked a tube from one ear, then the other. The fallen man arose, apparently restored to strength, and returned the wires to Pangar.

I noticed the next man. His entire breast had been torn away by the claws of the ramph. There was a set expression on his features, as of death or deep hypnotic sleep. But around the jagged wound was no sign of blood. The flesh, if it was flesh, was a peculiar grayish–red shade. And where the viscera would have been exposed in a normal human being, I saw a conglomeration of coils, tubes, wheels and wires, tangled and broken.

Pangar passed him by with but a single glance. "No use to try to save this one."

He rapidly examined the other fallen men. Two were picked up and slung over the shoulders of comrades. The rest were stripped of their weapons and helmets and left lying on the ground. A half dozen men, using their keen knives, had already skinned the ramph. It seemed that they wanted the hide only, not the flesh, for the great red carcass was left lying near the broken figures of the fallen men when we went.

Men or machines—which? I pondered the matter as Loralie and I walked beside the courteous and seemingly human Pangar, while the kroger waddled at our heels.

After a walk of about two hours we reached the summit of the mountain range and halted there for a few moments of rest while Pangar pointed with pride to the various features of the fertile valley of Doravia which was spread before us. It was oval in form, about twenty-five miles in length, tapering down to points at both ends where the inclosing mountain ranges ran together.

At the northwestern end of the valley a tremendous water fall, over a mile in height and fully a half mile in width, tumbled into a spray-veiled lake. From this flowed a river that wound through the center of the valley, to emerge at the southeast end. According to Pangar, it emptied into the Ropok.

At each side of the falls a conical, hive-shaped structure of immense size towered for some distance above the upper water level. These two enormous buildings were connected by an arched span that was fully a half mile above the lower water level. Their bases were hidden by the mists that arose from the bottom of the cataract.

The banks of the river, as it wound through the valley, were dotted at regular intervals by smaller twin towers of similar construction. The surfaces of all these buildings glistened with mirrorlike brightness.

In the very center of the valley, on an island of considerable size around which the river flowed in two nearly equally divided streams, was the largest structure of all. Cone-shaped like the others, but much larger than any of them, it reared its pointed, gleaming top to a height of fully two miles.

"The imperial palace of Tandor of Doravia," explained Pangar as he saw me looking at it. "A wonderful building. We will be there in a short time now."

"But it's fully five kants from here," I said. Then I noticed something which had previously escaped my observation. A thin cable stretching beside a long narrow platform a short distance below us extended out toward the tower, though it soon dwindled into invisibility. It was composed of the same peculiar glistening material.

"I have signaled for a car," said Pangar. "It will be here soon."

As I watched, a tiny gleaming speck became visible far out over the valley. Its apparent size grew larger with amazing rapidity, and in a few seconds I saw that it was a long, octagonal vehicle, pointed at each end, and constructed of the shimmering, transparent material.

It came to a stop beside the narrow landing platform without any perceptible jar or sound, and we all hurried down to meet it. When we reached the platform I found that round doors, hinged above, had been thrown open along the entire length of the vehicle.

Into one of these the princess and I were ushered by Pangar. The small kroger had kept close at our heels. We had no more than taken the comfortable springy seats when the doors clamped shut; the kroger was left alone on the platform, and we never saw it again—to my relief. The car then started smoothly out over the valley. In a moment it was speeding so rapidly that the landscape, though far below us, became a mere blur.

It seemed that only a few seconds elapsed before the car slowed down once more and we were entering an octagonal opening in the enormous central tower I had previously noticed. Before we entered I had a brief view of hundreds of other similar openings in the tower from which slender, transparent cables radiated in all directions.

The door snapped open, and as we stepped out on the landing floor Pangar said, "I will conduct you immediately to our Torrogo, as he wishes to greet you in person."

"How do you know that?" I asked, puzzled.

"His majesty instantly communicates his wishes by thought–transference to any of his subjects."

"Then you communicate with each other here by telepathy?"

"Not with each other," he replied, "except through our Torrogo or a member of the Committee of Twelve—kings who are thought—censors for the emperor. If I wish to communicate with a distant comrade, I send my thought to the member of the committee whose duty it is to watch over my mind. He receives the message and, if he approves, transfers it to my comrade or to the Torrogo."

As he talked, Pangar led us through a maze of hallways, the decorated floors, walls and ceilings of which were all of the same glasslike substance, but opalescent, so that, with light coming from all directions, we moved without casting shadows. It gave me a queer sense of unreality —as if I were moving in a dream from which I should presently awaken.

But when we were suddenly ushered into a huge and magnificent throne room, the many octagonal doors of which were guarded by warriors with drawn swords, the ceiling of which was fully a mile above our heads reaching to the very peak of the hive–shaped building, and my eyes beheld for the first time the grandeur of the Imperial Court of Doravia, I felt positive that only in a dream could such splendor have existence. I pinched myself repeatedly to make sure that I was awake.

My illusion of unreality, however, was instantly dispelled as we were led before the throne. Seated on its scarlet cushions was a powerful and commanding figure of a man. His high forehead and heavy eyebrows, joined at the center, reminded me of Dr. Morgan, but there the resemblance ceased.

The nose was Grecian rather than Roman in type, and the clean–cut features had the pale beauty of chiseled marble. It was a face which showed remarkable intellectual power and, at the same time, an utter lack of all sentiment or human sympathy. Although every other man belonging to this strange race was beardless, the ruling monarch wore, at the end of his chin, a narrow, sickle–shaped beard which curved outward and upward, ending in a sharp point.

Flanking each side of the throne was a row of six lesser thrones, on each of which sat a scarlet–decked individual whose insignia proclaimed the rank of rogo, or king. These rogos, I judged, must comprise the Committee of Twelve referred to by Pangar. On still lower thrones sat the purple–decked nobles of the land, while lining the walls on either side stood the blue–decked plebeians. Beyond these, on the outskirts of the throne, as it were, were massed a few of the gray–decked slaves.

Tandor stood up as we were brought before his throne—a deference due visiting royalty—and smiled, his black eyes boring into mine as we exchanged salutations. Although his smile was friendly, there was something about the look of his eyes which was not quite human. They appeared snakelike, with a sinister, hypnotic quality that was far from reassuring.

"You find me in the midst of my multifarious court duties," said Tandor, still smiling, "but I shall terminate them as soon as possible. Meanwhile, permit me to offer you rest and refreshment. Pangar will show you to the quarters provided for your entertainment. I shall join you presently."

When we were outside the throne room, Pangar issued instructions to a page, who hurried away, to meet us again down the corridor with a girl who wore the scarlet insignia of royalty, followed by the others whose purple ornaments proclaimed them daughters of the nobility. The six girls were shapely and quite pretty, but their mistress was beautiful. With a superb figure, glossy black hair and big black eyes, half veiled with long dark lashes, she rivaled the beauty of Loralie herself.

Yet, on comparing the two I was struck by a marked contrast between them. While the Princess of Tyrhana was the spiritual type of beauty, her every lineament suggesting purity and strength of character, this royal girl of Doravia appeared voluptuous, sensuous and apparently with great strength of purpose—like an exalted odalisque, or perhaps a fallen houri.

According us the royal salute, to which we responded in kind, she spoke softly with a low musical voice that, while it betokened culture and refinement, yet had about it a certain husky undertone which was puzzling. Her black eyes, too, I thought had something of that reptilian quality which had shone forth from the orbs of Tandor.

"I am Xunia of Doravia," she said. "It is the wish of my brother, Torrogo Tandor, that Loralie of Tyrhana be entertained in my apartments until such time as suitable quarters can be prepared for her."

She held out her hand to Loralie, who took it without hesitation, and the two moved off down a transverse corridor followed by the six handmaidens. Pangar then conducted me to a luxurious suite, whose glasslike furniture was upholstered with chlorophyl green ramph hide tanned to a softness that was almost velvety.

After a bath and a shave I felt greatly refreshed.

"His majesty is now ready to receive you in his private dining room," Pangar then told me.

A short walk down the corridor brought me to a doorway, octagonal in form, before which two guards stood, sword in hand. At a sign from Pangar they drew back two scarlet curtains, and I entered the room. As the curtains dropped into place behind me I beheld my royal host seated at an octagonal–topped table of translucent scarlet material in a high–backed golden chair upholstered with ramph hide, which was also stained a brilliant scarlet. He arose as I entered and tendered me the royal salute, which I returned. Then I took a chair at his right which an unobtrusive servant placed for me.

"I trust that you will pardon the slimness and coarseness of the fare which I am about to place before you," said Tandor after I had taken my seat, "but, with the exception of the slaves, we of Doravia do not eat or drink as you do in the outer world."

A slave set a crystal bowl before each of us. Mine was filled with steaming kova, but that which was placed before the Torrogo contained a heavier liquid which seemed to fume rather than to steam. It had an acrid smell which reminded me of the odor of a corrosive acid.

"May your years be as many as the stars," pledged Tandor as he raised his bowl to his lips.

"And may yours be as numerous as the rain drops that fall on all Zarovia," I replied, tossing off a draught of kova.

"Your arrival, O Prince," said Tandor, setting down his bowl, "was timed most opportunely, as you will realize from what I am about to relate to you. For the past two thousand years I have been planning a great experiment—one which if successful will revolutionize the lives both of my kind and yours."

"That is indeed interesting," I replied as a platter of chopped mushrooms and grilled ramph steak was set before me. "But—two thousand years?"

A disk–shaped vessel, black in color, was set before Tandor. Coiled about the handles on each side of the vessel were two insulated wires with electrodes on the ends. Uncoiling them, he inserted an electrode in each ear.

"I was born five thousand years ago in your country of Olba," he said, "the second son of the Torrogo. I did not covet the throne, preferring scientific research in chemistry, physics and psychology. When I had learned everything the greatest scientists of my time could teach me about these subjects, I began to combine my knowledge of the three with a view to realizing a dream of mine which is perhaps the universal dream of mankind—immortality.

"As I look back on my earlier efforts I realize how exceedingly crude they were, but alter countless experiments and untiring efforts, they worked. No doubt you have noticed the great difference between yourself and my people—between my sister Xunia and Princess Loralie."

"I saw the chest of one of your men, which had been torn open by a ramph," I replied, "and he was evidently no ordinary human being. I also heard talk of depleted power units, and I have noticed that you drink a beverage which smells and looks like fuming acid and that your food is evidently transmitted to you in the form of fluid power."

"In other words," said Tandor, "you have deduced that we are a race of automatons—machine men. You are right, but I do not believe that there exists anywhere else on any world a race of man—created beings with souls. Nearly five thousand years have elapsed since I cast off forever the frail shell with which nature endowed me to take up my existence in a more enduring body of my own creation.

"You are of course familiar with the phenomena of personality exchange and telekinesis. You are aware that two men can permanently or temporarily exchange their physical bodies.

"My problem, then, was to construct a duplicate material body into which my personality could enter, and which would respond to the direction of my will by amplifying the power of telekinesis. The first body which I succeeded in so entering collapsed because of faulty construction, and I barely got back to my own body in time to save it from dissolution and myself from being projected into the great unknown. But I made many others, and when they were at last perfected, I published my discovery in the Empire of Olba.

"My father had been received into the mercy of Thorth in the meantime, and my brother had succeeded him to the throne. I called on him to join me in immortality, and offered to make every person in the empire an immortal. To my great surprise and disappointment, my offer not only met with rebuff, but a systematized persecution against me and my followers was begun by the more religious of the Thorthans.

"Influenced by the religious leaders, my brother presently ordered the banishment of myself and my followers, who remained faithful to me. With less than a thousand of these I came to these shores and subsequent explorations revealed this valley."

I murmured my astonishment at all this.

"The only member of my family to accompany me," he went on, "was my sister, Xunia, who had been in sympathy with my plans from the first. As rapidly as I could, I prepared duplicate bodies for my followers, it being necessary to give each body the outward semblance of the body and brain which was to be quitted, else the personality would not enter it.

"I have always kept many bodies in reserve for myself and for my sister, so we were prepared for almost any emergency. In case the body I occupied broke down I could instantly enter another. If that one broke down or was destroyed, I could enter still another, and so on.

"The slaves were the only class which was never completely immortalized. Today, immortalization of a slave is a reward for faithful service. You may readily see, therefore, why the food and drink for which I am forced to apologize are of the cruder sort. I am compelled, for the moment, to offer you but the fare of slaves."

"It is excellent," I replied, "and quite good enough for any king's son."

"I will find the means to improve it, however, as I expect you to remain here permanently. I have planned a great honor for you."

"Indeed?"

"I will explain. As you probably have surmised, there has been no such thing as propagation of the race among my immortals. This did not bother me in a material way. When I lost a follower—which was rarely, as every one has at least one extra body and most of them several—I could immediately replace him from the ranks of my slaves. But there was no love; and after about three thousand years had passed, the defect bothered me emotionally.

"I knew that the problem which confronted me was considerably more difficult than any on which I had previously worked, but undaunted, I plunged into my studies. Two thousand years of anatomical, histological, embryological, biological, biochemical and psychological research have brought their reward, so that, although today I differ from you physically as much as ever, I have built into my newest bodies and into those of my sister the sexual characteristics of ordinary human beings.

"Pangar was sent forth today with the object of bringing me two human beings suitable for marriage with royalty. His journey ended almost as soon as it began when he found you and the princess. I therefore offer you the hand of my beloved sister in marriage, and will likewise offer the half of my throne to the Princess Loralie."

"But if we should decline the honor?"

"It is unthinkable. Even if you were to decline, either of you, I have means at hand which, I am sure, will cause you to reconsider gladly."

Removing the electrodes from his ears and draining his bowl, he arose and summoned two pages. To the first, he said, "Instruct the Princess Loralie to prepare for my coming." As the messenger sped away he said to the other, "You will conduct His Highness Torrogi Zinlo of Olba to the apartments of Her Highness Xunia, Torrogina of Doravia."

As the little page conducted me to the apartments of Princess Xunia I turned over in my mind Tandor's strange story and its revolting sequel. I was going to the apartments of a girl who had been dead five thousand years, but whose soul was bound in a machine. Beautifully and cleverly constructed as it was, it was yet a mere mechanical contrivance—a thing of wheels and cogs, levers and shafts, a thing that fed on electrical energy and drank fuming acid.

And I was expected—commanded with a none–too–veiled threat—to make love to this travesty on life.

But Loralie! Somehow I must contrive to live in order to save her.

The page stopped before an ornate doorway, two guards saluted and opened massive doors. Then a pair of scarlet curtains were drawn back, revealing a luxurious boudoir. "His Highness, Zinlo of Olba," announced the page as I entered the room.

The curtains fell in place behind me. I heard the guards close the heavy doors.

As I looked at the beauteous dead—alive creature that reclined on a luxuriously cushioned divan in a scarlet and gold decked recess, a feeling of revulsion swept over me; yet, paradoxically enough, this was combined with admiration. I was revolted at thought of the nearness of this living dead thing, but could not but admire the consummate art that had created so glorious an imitation of the human form.

I realized that if I would live to be of assistance to Loralie I had a part to play.

Xunia smiled languidly, seductively, as I stood before the raised divan just outside the niche it occupied. With feline grace she extended a slender, dimpled hand. Shuddering inwardly, I took it, expecting to feel the cold clamminess of death. But it was as warm as my own and as natural—from its white back in which a delicate tracery of blue veins showed, to the pink–tipped, tapering fingers. I raised it to my lips and released it, but she clung to my fingers for a moment, pulling me to a seat on a low ottoman just in front of her.

"Long have I awaited your coming, prince of my heart," she said. "Be not afraid to come near to me, for it is my desire and my command."

"To be prince of your heart were indeed an honor," I replied, "yet you name me this, having only seen me today."

"The moment I saw you I knew it was so. Fear not, beloved, that there have been others before you. I am, and have ever been, virgin in mind as in body. Once I thought I loved, yes, but it was long ago, and then I was but a child."

"You make me very jealous, nevertheless," I said, remembering the part I had to play.

"I did not really love him, I swear it, dearest." She ran her fingers through my hair in a gentle caress so natural, so womanly, that I found it well–nigh impossible to believe her other than a real princess of flesh and blood.

Then, before I realized what she was about, she twined her arms about my neck and kissed me full upon my lips.

The kiss did not taste of acid, as I had imagined it would, but was like that of a normal, healthy girl, though it aroused in me a feeling of revulsion which I was at some pains to conceal.

"I go now, beloved, to prepare for your marriage," she said. "Await me here."

As I stood up, she took my hand and arose gracefully. The time for action had arrived. Yet, as I looked down at the slender, beautiful figure, the long–lashed eyes gazing trustfully up into mine, I hesitated to carry out the plan which I had been contemplating as I sat there on the ottoman before her—a plan with which I hoped to accomplish a double purpose—to rid myself of this machine–monster and to get her brother away from Loralie, for she would probably summon him telepathically, if in no other way.

I was trying to think of her as a dead thing in a machine, yet it seemed impossible that she was other than human, so natural was she, and so beautiful. But the thought of Loralie and the danger she was in steeled me to the task.

Seizing Xunia by her long black hair, I whipped out my stone knife and slashed the artificial muscles of the slim white throat. She gave one startled scream, which ended at the second slash of my knife, and went limp as I jerked the head backward, cracking the metallic structure which took the place of cervical vertebrae. Instead of blood, there spurted from the severed neck a tiny stream of clear fuming liquid, a few drops of which fell on my hand, burning it like molten metal.

Dropping the sagging body, I turned and was about to part the curtains which led out into the hall to see if the coast was clear, when I heard a stealthy sound behind me. Swiftly turning, I saw Xunia, apparently unharmed. In her right hand was a long, straight–bladed sword drawn back for a thrust. Behind her lay the body I had just destroyed.

I leaped back just in time to avoid her vicious lunge. Then, jerking my spiked club from my belt, I dealt her a blow which crushed her skull like an egg–shell. But scarcely had this body sunk to the floor ere a panel opened in the wall behind it and a third, armed like the second, stepped out to attack me.

"Fool," mouthed the advancing figure. "Think you that you can slay one of the immortals?"

This time she swung the sword with both hands with the evident intention of decapitating me, but I struck the weapon from her hands. Then I crushed the skull of this third body.

I leaped through the opened panel, where four more bodies, identical with the other three, lay on scarlet couches. The one nearest me was just sitting up, when I smashed the skull with my club. I quickly disposed of the next two in the same manner before they showed any signs of life, but the last rolled from the couch and, dodging beneath my arm, rushed out into the room from which I had just come.

"Brother!" she screamed. "Brother—he would destroy me!"

As I stopped the screeching of this last figure with a blow of my club, the entire wall toward which I was facing rolled up like a curtain. On the other side of it was a room like the one in which I stood, and in that room were Loralie and Tandor.

The long hair of my princess was disheveled and her eyes were flashing with anger as she tried to pull away from the monarch, who gripped her slender wrists.

Taking in the situation at a glance, Tandor suddenly released Loralie, who fell to the floor. Then he whipped out his sword and advanced on me.

Forgetting that I held only a wooden club, I bounded forward to meet him. A sneer crossed his cold, statuesque features, as with a deft slash he cut my club in two near the handle.

"Die, upstart," he snarled, raising his weapon for the blow that was to end my existence.

I barely succeeded in avoiding death by leaping back, then caught up one of the swords which Xunia had dropped.

But as I attacked he came on guard and countered with a skill which spoke of expert training and thousands of years of practice.

"In your ignorant folly," he said, cutting, thrusting and parrying with a deft precision which amazed me, "you believe you have sent my sister into the unknown, and that with your skill as a swordsman you can do likewise for me. Know, then, witless one, who would try conclusions with the immortals, that in one of the great twin towers which flank the falls under constant guard, my sister has twelve more bodies in reserve.

"Should you succeed in destroying the six bodies I have here in the palace —which you will not be able to do—I also have twelve more under guard in the opposite tower."

"I care not if you have a hundred, you monster," I retorted. "Bring them one by one within reach of my blade and I'll eventually send you down the unmarked trail you should have taken five thousand years ago."

"You are, I perceive, a braggart as well as a dullard," said Tandor. "You realize, of course, that I can call the guard and have you slain at any moment I choose to do so. Yet to make things more interesting I'll make a wager with you. If you succeed in besting me and destroying the six bodies I have here in the palace, I'll promise not to alarm the guard until I return from the tower in one of my reserve bodies. If I force you to surrender, you are to become my slave for life, body and soul, to do with as I see fit. Is it agreed?"

"It is a wager," I replied between clenched teeth as I desperately sought for an opening in this, the most marvelous guard I had ever encountered.

Tandor laughed as I tried, one after another, the many tricks I had learned in my fencing on three planets.

"You are a good swordsman, youth, better than any mortal I have ever encountered; yet I, with five thousand years of training, am merely playing with you. See, I can touch you at will."

And with that, he pinked my left shoulder.

The moment he extended his weapon he left the opening for which I had been waiting. Not knowing on what part of his anatomy I could use my point effectively, I dealt him a swift neck cut with its keen edge.

The head flew from his shoulders and bounded to the floor, but the body did not fall. Instead, it stooped, and catching up the head, tucked it under its left arm and resumed the contest. Here, indeed, was a super–mind, which could control, at the same time, severed head and body.

"A pretty counter," mocked the head, while our blades clashed as vigorously as before, "but perhaps not as effective as you expected. I will tire you out presently. Then will I slice you down, inch by inch, until you will be glad to yield."

"Not with this body," I replied as I got inside his guard for a swift downward cut on his forearm. Cleanly severed, it fell to the floor, the hand still gripping the sword. An instant later the body dropped the head and fell. Then a panel slid up behind it, and Tandor, another sword in hand, emerged, smiling sardonically. "You are more clever than I thought, princeling, but that trick will not work again."

"It is not the only one I know," I replied and, catching his blade on mine, disarmed him, much to his consternation. This time I not only split his head from crown to chin, but slashed off his right arm. Then I rushed through the panel opening in time to catch a third newly animated body just arising from its scarlet couch. I served it in like manner, but the fourth sprang up before I could strike and came on guard with appalling swiftness. Before Tandor could attack in this body I struck two swift blows, splitting the heads of the two recumbent forms.

I stepped to one side barely in time to avoid a powerful downward cut that would have divided my own head had it landed, and before he could recover I severed the sword arm of my attacker and split his head.

Rushing back into the room where I had left Loralie, I found her plucking a sword and dagger from one of Tandor's bodies.

"We must get out of here at once," I said. "In a few moments Tandor will be back here in one of his swift vehicles. Then, the terms of the wager fulfilled, he can quickly have us captured."

"But where can we go? How can we possibly escape?"

"I do not know, but we most certainly can't get away by remaining here. Come."

Chapter XI

CAUTIOUSLY PARTING the scarlet drapes which hid the doorway, I saw that the heavy doors had been bolted. Tandor had evidently intended that he should not be disturbed.

I expected that there would be guards in the corridor, and therefore decided that a bold front would serve our purpose the best. I appropriated one of Tandor's magnificent belts with ornate sword and dagger, and outfitted Loralie likewise with one of Xunia's belts which contained lighter weapons. Then we walked quietly to the doors, which I unbolted and swung back. The guards saluted stiffly and closed them after us as we passed out.

"It is the command of his majesty," I said, "that he be not disturbed by messengers or others."

"To hear is to obey," replied both guards in unison as we strolled away down the corridor.

I only knew my way to one part of the building—the landing floor. After threading so many hallways, passageways and ramps that I had begun to think I had lost my way, we came out on the central landing platform, from which radiated the cables that carried the swift-moving octagonal cars to the various power houses of Doravia.

Glancing in the direction of the twin towers, I saw a car swiftly approaching from each and surmised that Xunia and Tandor were already on the way to the palace.

"Quick!" I said to Loralie. "We have not a moment to lose!"

Hurrying her to the side of a car which hung on a cable that pointed toward the south, I helped her aboard—then spoke to the pilot. "It is the desire of his majesty the Torrogo that we inspect some of the buildings of Doravia. You will first take us to the power plant at the southernmost end of the valley."

He saluted respectfully, then moved a control lever. The doors closed and we glided smoothly away from the platform. In a moment we were speeding swiftly southward at a dizzy height above the valley.

One by one we sped past the towers which dotted the river bank, so swiftly that each washed for but an instant in our range of vision. Yet it seemed to me that our pace was exasperatingly slow, for I knew that Tandor would surely reach the central tower before we arrived at our destination; if he made inquiry at the landing platform he would flash a message to the commander of the southern tower, and we would face arrest as soon as we arrived.

I accordingly loosened my blade in its scabbard and spoke softly to Loralie. "We must be ready to make a dash for it as soon as the doors open. Keep behind me, and I'll try to cut a way through."

As we drew up to the landing platform I saw a score of guards lined up to meet us. In front of them stood a captain with drawn sword.

The doors opened and we stepped out.

"By order of His Majesty..." began the officer.

I did not wait for him to finish but whipped out my sword and beheaded him before he could say more. Then I sprang forward and cut my way through the line of surprised guardsmen with Loralie close behind me. She drew her own weapon, and used it with more skill than I had believed possible in a woman.

As we dashed off down a corridor we met two more guards, but they were crude swordsmen and detained us but for a moment. On coming to a transverse corridor, we turned, hoping thus to elude our pursuers; but a moment later they rounded the turn, and at the same time I saw a large party of men closing in on us from the opposite direction.

"We're trapped," I said, "and this is a poor place to make a stand. We'll turn in at the next doorway we come to."

There were doors on both sides of the corridor at intervals of about fifty feet, and I accordingly stopped at the next and wrenched it open. Without looking to see what was within, I pushed my companion into the opening.

Hearing a scream and a thud, I leaped in after her, but scarcely had I slammed the door ere my feet slipped from under me, and, half lying, half sitting, I found myself sliding down a steep spiral incline in total darkness at a terrific rate of speed.

For several minutes I continued my downward course uninterrupted. Then the incline grew less steep and I glided over a series of humps which retarded my progress. A moment later I shot out into the air and daylight, my feet struck a cushioned wall, and I fell on a thickly padded floor.

Springing to my feet, I saw Loralie standing with drawn sword, facing a huge guard. A short distance behind him wavelets from the river lapped the edge of the floor on which a half–dozen narrow, pointed boats made from the transparent metal were moored.

As I dashed forward, the guard struck her sword from her hand and attempted to seize the princess, but ere he could do so I sprang between them and our blades met. Aside from Tandor himself, he was the cleverest swordsman I had encountered in Doravia.

Back and forth we fought on that moist, slippery floor, until I succeeded in forcing him to the water's edge. Binding his blade with my own, I pushed it upward, and leaping in close, struck him in the breast with my left fist. He toppled for a moment on the brink—then fell into the river behind and sank out of sight.

At this instant I heard the clank of arms in the chute behind us, followed by the thud of a body against the padded walls, then another and another.

Quickly flinging Loralie into one of the boats, I slid it to the water's edge, leaped in and shoved off. Four spadelike paddles lay in the bottom, and seizing one of these I managed to get several boat lengths from the shore before our pursuers reached the water's edge.

The first boatload was not long in putting off after us, and with four paddles working it gained on us rapidly. Behind it, another and another left the shore until five in all pursued us.

Seeing that it would be only a few moments before we were overhauled, I strung my bow and shot an arrow at the foremost paddler. Although it pierced his breast it did not seem to discommode him in any way. He paddled forward as briskly as ever, pausing only to snap off the shaft and fling it into the water. I tried a second shot, this time aiming for his head, but the arrow glanced harmlessly off his glittering, transparent helmet.

Loralie, following my example, also strung her bow and tried a shot at the second paddler. It struck him in the arm, but he broke off the shaft and continued his paddling as if nothing had struck him.

"Save your arrows," I said as a plan suddenly occurred to me. Quickly unwinding a length of the cord I still had with me, I looped part of it and cut it in short pieces. Then I took from the ammunition belt of Talibot a clip marked "Tork Projectiles—Explosive." Extracting one, I bound it to the head of an arrow and discharged it at the first paddler. He grinned derisively as he saw me raise my bow, but his grin disappeared, together with most of the upper part of his mechanical anatomy when the missile exploded.

Passing several projectiles and bits of string to Loralie, I quickly prepared another arrow and blew a second pursuer out of existence. By this time the first boat was less than thirty feet from us, and I knew I would not have time to prepare a third arrow, so I drew my sword and made ready for the attack of the two guardsmen who remained in this boat. But before they came alongside there was only one, as Loralie, having prepared one arrow, proceeded to blow the other to bits.

The last remaining guardsman leaped to his feet as the slender prow of his boat struck the rear of ours. Dropping my sword in the bottom of our boat, I quickly tipped his boat to one side. The fellow tried to maintain his balance by throwing his weight in the opposite direction but I had anticipated this, and as he did so I reversed the tilt of his boat, precipitating him into the water where he sank out of sight.

So occupied had I been with our pursuers that I had not noticed whither the swift current was carrying us. My first intimation of danger from this source was a bump and a grinding noise as our keel struck and then slid over a submerged rock, nearly capsizing us. I seized a paddle and swung our craft parallel with the current just as we were precipitated into a seething, whirling rapids, from the foaming surface of which projected numerous jagged rocks.

I bent all my efforts to the task of avoiding the dangerous rocks which loomed ahead as we shot forward with alarming speed, now on the crest of a huge wave, now in a hollow so deep we could not see out of it. As we advanced the river became narrower, the rapids steeper, and the rocks more menacing. It appeared that the River of Life—for such Pangar had named it to me—might become, for us, the River of Death.

Try as I would, I could not keep our craft from repeatedly colliding with the rough boulders that now beset our path. The strength of its transparent metal sides astonished me.

We were nearly through the rapids, and I was just breathing a sigh of relief, when the unexpected happened. Our prow struck a hidden point of rock, the boat swung broadside, and we turned over.

I heard a scream from Loralie as I plunged into the water, head first. The metal paddle to which I had unconsciously clung as I fell quickly carried me to the jagged bottom. I let go and swam as rapidly as I could to the surface. Shaking the water from my eyes I looked around. The swift current had already taken me beyond the foot of the rapids into deeper water. I could see no sign of the princess, though I craned my neck in every direction.

Our overturned boat had drifted past me, and three more boats were swiftly descending the rapids, bottom up, but behind them came two more, in each of which sat four Doravian guardsmen.

Filling my lungs, I dived for the spot where I thought Loralie might be, and swam under water for some distance.

Upon again coming to the surface, I saw her swimming for the shore about a hundred feet ahead of me. Our drifting boat had hidden her from my view.

I saw the first boatload of Doravians pass the bottom of the rapids unscathed as I struck out after the princess. But as soon as they reached calmer water they plied their paddles with such dexterity that I knew they would overtake me long before I could reach the shore.

Although I was greatly hampered by the weight of my weapons, I hesitated to part with them, since I could not possibly get to land ahead of that boat, even if I were stripped.

Presently the boat came within fifteen feet of me. The foremost guardsman laid down his paddle and drew his sword. Raising the weapon above his head, he leaned out over the bow to dispatch me. At this instant I dived, and describing a loop under water, came up just under the stern of the boat. Seizing it in both hands, I capsized the craft, plunging my four assailants into the water. None of them reappeared. The metal men apparently could not swim.

By this time the last boat had negotiated the rapids and was paddling swiftly toward me. Again I struck out for land, this time with some hope of making it. Loralie, who had just reached the shore, called out to me, "Hurry. A silticum is coming this way."

I looked back, and my first view of a silticum was none too reassuring. It was an enormous reptile with a green lizardlike body, serpentine neck, and a head of immense proportions.

I struck out desperately for the shore, and the paddlers increased their efforts. The noise they made attracted the attention of the reptile. Suddenly swerving, it made for the boat.

As I was quite near the shore I lowered a foot, struck bottom, and waded out just as I stepped on the sloping beach, an exclamation from the princess made me turn.

With serpentine neck arched and mighty jaws distended, the huge saurian lunged downward, straight for the center of the boat. One of the occupants rammed his sword in that cavernous maw, and two others slashed at the scaly neck, but with no apparent effect on the reptile. It seized the boat in its immense jaws and lifting it high out of the water, shook it as a terrier shakes a rat. Hurtling through the air to the right and left, the bodies of the four Doravians fell into the river and disappeared.

"Come," said Loralie, tugging at me arm. "That creature is as swift on land as in the water. Let us get out of its sight before it takes a notion to follow us."

"With pleasure," I responded, and together we hurried up the bank and plunged into the fern forest.

For some time we ran forward, side by side, sinking ankle–deep in the soft moss that carpeted the forest floor.

"I'm thirsty," said Loralie, "and hungry. Aren't you?"

"Ravenous. Nothing will satisfy me but a good big steak. Spore pods are all right for appetizers, but to satisfy hunger there is nothing like meat."

"I've lost my bow and arrows," she said, ruefully, "along with that clip of explosive projectiles you gave me. I dropped everything when the boat tipped over."

"Never mind. I still have my bow, plenty of arrows, and another clip of explosive projectiles. It's a man's place to bring in the game, anyway, while the woman looks after the home."

"The home? What do you mean?"

"Why—er—that is, I was just drawing a comparison between ourselves and primitive people. The man went hunting, you know, while his mate looked after the cave, or tree, or whatever they lived in."

"His mate? I fail to see the comparison."

"Well, you know we're leading a rather primitive existence just now, and...

"Prince Zinlo," she said, suddenly stopping and facing me, "will you cease talking generalities and tell me just what you mean?"

"Yes," I cried vehemently. "I'll tell you what I mean. I hadn't intended to, but it seems my words betray my thoughts. I love you, Loralie. I want you for my mate—my princess. But as you so plainly dislike me I shall probably go on desiring you until the destroyer of all desires puts an end to my existence."

"I was beginning to wonder," she said softly, "if I would ever get you to say it."

Before I realized the purport of her words her arms were around my neck—her warm red lips upturned, inviting. I crushed her to me, and found her a new Loralie—tender, yielding, passionate.

"I've loved you since the very hour we met," she said, "when you tossed my presuming cousin into the shrubbery."

Her hand caressed my cheek, roving softly over my rugged face. But as I bent to claim the sweetness of her lips, I heard a twig crack behind me, and I whirled about, hand on hilt.

To my amazement I beheld Prince Gadrimel, standing only a short distance from us. "A thousand pardons for this intrusion," he lisped. "By the beard of Thorth, I could not find the heart to disturb so pretty a love scene, were it not that darkness approaches and the camp is a considerable journey from here."

Too astonished to reply, I could only stare at him as he stood with a mocking smile on his effeminate features, toying with a jeweled pendant on his breast and ogling Loralie.

"No doubt you are glad to see me, fair cousin," he continued in his mincing patoa, grinning at the princess, "so glad that the joy of my coming overwhelms you—renders you speechless. Come, haven't you at least a little cousinly kiss for your deliverer who has come so far to rescue you? You appear to lavish your caresses quite generously outside the family."

My blood boiled at his studied insolence, his air of proprietorship, yet I strove to control my feelings as I answered him. "The kisses of the Princess Loralie are her own to bestow. You will do well to remember that, Prince Gadrimel."

"And you, Prince Zinlo, will do well to speak only when spoken to." Gadrimel held out a hand to Loralie. "Come, cousin, let us get to camp before darkness falls. By tomorrow we will be aboard my flagship and well on our way to my father's palace."

The princess drew closer to me and looked up into my face as she answered, "Prince Zinlo is my fiance. I'll go where he goes."

"This nonsense has gone far enough," said Gadrimel, sharply. "Ho, warriors!"

Scarcely had he uttered his call ere there closed in on us from the surrounding fern brakes a full hundred armed men of Adonijar.

"Seize and bind this interloper," he commanded, pointing to me.

When this had been done, Gadrimel stationed a stalwart soldier at my side. "Remain here with the prisoner, until we have passed out of earshot. Then..." He stepped close to the soldier and whispered something to him. "For which," he concluded, as he stepped back, "you may have his weapons, accouterments and anything else of value he may have with him."

Loralie attempted to come to me as I stood there, bound hand and foot, but two soldiers prevented her.

"What are you going to do to him?" she cried.

"Now, now. Calm yourself, sweet cousin," said Gadrimel. "I am but sending him on a journey. I must insist that you hurry to camp with me at once, or darkness will overtake us on the way; the night–roving beasts will not be pleasant to meet in this forest."

In spite of her struggles he dragged her away. Behind them moved the entire company of warriors with the single exception of the one who had been instructed to remain with me. He stood immobile, listening until the sound of voices and the clank of weapons had died away in the distance. Then he turned to me.

"I have been commanded to kill you, Highness," he said, simply. "Never before have I slain a bound and helpless man, but I am a soldier of Adonijar and may not disobey the command of my prince. However, I was not instructed as to how I should kill you, and I bear you no malice. By what weapon do you choose to die?"

"The sword," I replied, "has ever been my favorite weapon. If I must die now, let it be by the sword."

"The sword?" he asked in puzzlement.

"That long straight–bladed weapon in the sheath at my feet," I answered. "Plunge it into my heart and get it over quickly."

Slowly he bent over and withdrew the sword from its sheath. He examined it curiously, testing the sharpness of its point with his palm and the keenness of its edge with his thumb.

"By the blood of Thorth!" he exclaimed. "This is a beautiful weapon. And it will be mine as soon as I have slain you. Make ready, now, to die."

Chapter XII

As I STOOD there in the fern forest bound hand and foot and helplessly awaiting the death blow at the hands of Prince Gadrimel's henchman, I was suddenly knocked flat by the drop of a huge, furry body from the limbs of the tree above me. Half dazed, I sat up just in time to see a female cave–ape crush the head of my would–be slayer with her sawedged club.

She turned, and as she did so, I recognized her features.

"Chixa!" I exclaimed.

"Long have the cave–apes sought their Rogo," she said, "and great will be their rejoicing when he returns."

With her flint knife she quickly cut my bonds, and I stood erect once more, stamping my feet and chafing my wrists to restore circulation, scarcely able, as yet, to understand that I was really alive.

"Do you cave–apes still consider me their king?"

"According to the custom you would lose your kingdom if you remained away for more than one endir. But you have been gone only a few days. As there is much judging to be done, we have been searching for you."

"Where are the other searchers?" I asked.

"Many of them are within call."

"Then call them, and let them call as many others as they can."

With marvelous agility for a creature of such great size, she scampered up to the leaf crown of a tall tree–fern. Then, cupping her paws, she gave utterance to a queer, trilling cry. It was answered, not once, but many times, from various points far and near.

Then she descended the tree and dropped into the glade beside me.

Presently there came swinging through the branches a great, yellow-tusked male who, as soon as he saw me, roared, "Hail, Zinlo!" and dropped to the ground near me. Another emerged from the fern brakes, repeating the salute of the first, and it was not long before I was surrounded by more than two score males and about half as many females.

As these shaggy man-beasts sat grouped around me, respectfully waiting for me to speak, their demeanor showed that they recognized me as their king without question.

"My subjects," I said, "I have work for you in which there is much danger and much fighting."

"Will there be food-men?"

"There will be many food-men."

"Good!" This answer was unanimous.

"We will start as soon as I have issued full instructions."

But the great, yellow-tusked male who had first responded to the summons of Chixa protested, "There is judging to be done. Will you not first do the judging, so we may go into the fight with our differences settled?"

"Who are you," I asked, "to question the edicts of your Rogo?"

"I am Griff, mighty warrior, mighty hunter," he replied, puffing out his broad, hairy chest. "But I do not question your edicts. I only ask that you hold the judging now."

Before I could answer him there came a sharp cry from a female who had perched herself in the branches above our heads in order that she might better observe everything that went on.

"Danger! Danger!" she shrieked. "A silticum!"

Every cave-ape instantly took to the trees, and I heard the crashing of a huge creature in the underbrush as it swiftly made its way through the forest. Evidently the silticum which had attacked the Doravian guards had seen us, even as Loralie had feared, and was now on our trail.

Quickly taking the last clip of explosive projectiles from my belt, I removed two of the needle–like missiles and bound each to the head of an arrow. Then I strung my bow and awaited the coming of the monster.

Chixa called to me from the leaf crown of a tall tree–fern. "Come up into the trees, Rogo. You cannot fight a silticum."

"Yes, climb before it is too late," called Griff. "No one has ever slain a silticum."

Although I knew nothing of the ways of this saurian, I had seen its great size and knew that if it had intelligence enough to do so it could pull down any tree within my range of vision. In view of this fact, and also because I could not get about as swiftly as the cave–apes in the trees, I felt safer on the ground.

"Stay up in the trees if you like," I answered them. "I will show you how your king slays a silticum."

In a few moments I saw the huge green head swaying on the snaky neck at a height of about twenty feet above the ground. It was looking this way and that, apparently searching for me. As it drew closer I saw that it was indeed the same monster that had attacked the machine men in the boat, for projecting through its lower jaw was the transparent sword blade where the Doravian guardsman had thrust it, and which the creature had been unable to dislodge.

I fitted an explosive arrow to my bowstring, and at this moment the monster spied me. With a hiss like steam escaping from a locomotive, it distended its enormous jaws and charged straight for me. Taking careful aim at the cavernous maw, I drew the arrow back to the head and let fly.

The reptile turned slightly so my shaft did not strike the target squarely, but considering the terrific force of the tork projectile this did not greatly matter. For although the missile struck the monster in the corner of the mouth, the explosion tore off the whole side of its head.

I instantly fitted my second arrow to the bowstring, but instead of advancing the great saurian swerved to one side and began threshing about in a circle, striking this way and that with its huge, scaly tail which swept the fern trunks before it, knocking them over as if they had been mere reeds. As the tail now appeared to be the most formidable weapon of the beast, I aimed my second shaft with a view to crippling this appendage, and let fly.

It struck the monster just above one of its thick hind legs, blasting a great hole in the flank and not only crippling the tail but both hind legs as well.

Upon seeing this, the cave–apes instantly descended on the stricken reptile with yells of triumph, and were soon hacking at its heaving sides with their saw–edged clubs and prying up huge scales with their flint knives in order to get at the quivering flesh underneath.

"Hail, Zinlo!" the shouted. "Mighty warrior, mighty hunter, mighty sorcerer! With his magic he slays even the silticum, the terror of stream and forest!"

As I watched the cave–apes at their bloody feast, I recalled that I, too, was hungry. Elbowing my way through the growling, snarling, milling mob, I carved a steak from the shoulder with my keen Doravian dagger. Then I made a small cooking fire and grilled my slab of meat. It proved tasty enough, although rather tougher than a gourmet would have relished. But with good teeth and an excellent appetite this bothered me not at all.

By the time I had finished, and swallowed a draught from a water fern, my hairy retainers had all gorged themselves.

I arose and called them together. They squatted expectantly around me in a semicircle. "You, Griff," I said. "Bring me that shiny club which sticks in the jaw of the silticum."

After he had brought me the sword of the Doravian boatman, I continued, "You have asked that judging be done before we fight. I have no time for judging now, so I am going to let you do it. This shiny club will be your token of authority, by which you will do judging in my name. Go now, taking the shes with you, back to the caves. And beware that your decisions are just ones, for I will hear of it, and will come and slay you with my magic if they are not."

"But Rogo," he protested, "I would like to go and fight the food–men with your others."

"You will do as you are bidden without further question. Throw away your old club and take this shiny one which slays with its point as well as its edges."

Silently, and rather sullenly, he removed his club from his belt string and tossed it away. Then he took the sword and lumbered away through the forest, followed by the females.

As soon as they had departed I called the others together and started off on the trail of Prince Gadrimel. But darkness overtook us before we had gone more than five miles, and we were forced to take to the trees to avoid the depredations of the night–roving carnivora.

Propped in a high leaf crown that swayed with each passing breeze I didn't get much sleep during that noisy night, oppressed by my constant fear for Loralie in the clutches of her unscrupulous cousin.

It was with a sigh of relief that I greeted the dawn and made my way to the ground. Impatient to be off, I stopped only for a drink of water, then started down the well–marked trail with my small but formidable company. The spoor of Loralie's abductors continued to follow the winding course of the River of Life for about six miles to the remains of a large camp which had been completely surrounded by watch fires. Most of these were still smoldering as we came up.

Of the people of Prince Gadrimel we saw no sign, save tracks leading to the river where there were indentations made by the prows of small craft.

I led my ape–men at a trot along the flat, sandy beach for miles. The river bank gradually grew more rugged, and at last we climbed to a rocky eminence commanding a view of both sea and river.

Anchored not more than an eighth of a mile off this point, and rocking in the gently rolling swell, I saw the five ships of Prince Gadrimel. Paddling swiftly toward them from the river mouth were a score of small boats, in the foremost of which were two scarlet–clad figures which I knew must be Gadrimel and Loralie.

Helplessly I watched while his henchmen bundled the princess aboard the flagship, boats were drawn up to their places on the decks, sails were hoisted, and anchors weighed.

So, with straining eyes, a great lump in my throat and a weight in my heart, I saw Gadrimel triumphantly sail away over the bounding, blue–gray Ropok with the only woman I have ever loved.

As I stood there, absent–mindedly watching my subjects scurry through the forest in search of game, I pondered my predicament. The only thing left for me to do, I reasoned, was to follow the coast northward as Loralie and I had planned to do. In order to reach Olba I would pass through Adonijar, but single–handed I could do nothing against an entire nation.

Once in Olba I felt that I could persuade the Torrogo to let his supposed son have an air fleet for the purpose of avenging the attempted murder of the Crown Prince, and with this I could quickly persuade the ruler of Adonijar to give up the princess.

I dreamed thus futilely until a great splash of rain struck me in the face, followed by the patter of many more on the leaves around me. Brought to a sudden realization of my surroundings, I noticed that the gentle wash of the waves against the shore had changed to the booming roar of huge breakers, that the trees were bending before a considerable breeze, and that despite the fact that the day was not yet spent it was growing steadily darker.

A terrific peal of thunder, followed by a vivid flash of lightning, made every cave–ape drop the bone he was gnawing and look toward me as if for protection or guidance.

"Zog makes magic in the heavens, Rogo," said Rorg, quaking with fear. "Zog is angry. Let us hide until he goes away. I noticed a great cave beneath the next cliff when I was hunting."

Glancing around at the other beast–men, I saw that Rorg was not the only one who had been frightened by the peal of thunder. Every cave–ape was shivering in abject terror.

"Lead the way to the cave, Rorg," I said. "I do not fear Zog, but there is much rain and much wind coming from across the big water, and a cave will be more comfortable."

The frightened cave–ape needed no urging, but hurried off at once, the others after him, while I brought up the rear at a more leisurely pace. Peal after peal of thunder sounded, the lightning flashed almost incessantly, and rain came down in torrents before I reached the cave mouth.

Entering, I beheld my erstwhile fearless fighters huddled together like frightened frellas and shivering as if with the ague.

"Every one fears Zog," explained a young ape.

"Your Rogo does not fear him," I said, "and you should not. Come and help me pile stones in the doorway lest a silticum or some other monster get in tonight."

"We are afraid to go to the doorway," quavered Rorg. 'Zog will slay us with his magic fire."

"Enough of this. Come over here and help me, every one of you, or I will slay you all with my magic."

The tragic fear which was in their eyes was pitiful to behold, but they were not long in choosing between what they believed would be sure death from my magic and possible death from the bolts of the deity they called "Zog." The doorway was soon so completely blocked that no night–roaming beast could enter.

Night having come on by this time, the only light in the cave was from the frequent flashes of lightning.

For a long time I stood at the entrance. Each lightning flash showed branches flying through the air, fern–trees blown over, and wild things, large and small, scurrying for shelter.

I was awakened in the morning by a loud clatter and the sound of gruff voices. Sitting up with a yawn, I stretched my cramped limbs as I watched Rorg and several other cave–apes dragging the barricade away from the cave entrance. Gone was the unreasoning fear that had gripped them the night before.

I rose and followed them outside. The storm had vanished, and other than the upper cloud envelope which is ever present in the Zarovian sky, the heavens were clear. But the still–dripping fern forest plainly showed the ravages of the tempest. The ground was littered with leaves and branches;

trees were bent over, snapped off and uprooted, and many streams of muddy water trickled riverward.

Crossing the gulch which separated our cave from the highest eminence, I climbed to the point where I had been standing the night before when the storm struck, to find some spore pods. As I gazed out over the Ropok, now rolling as gently as it had before the storm, my munching terminated in a sudden exclamation of surprise.

Lying on their sides far out in the surf with the waves rolling over them, and apparently deserted, I saw the battered hulls of two of Prince Gadrimel's ships. And anchored on the lee side of the promontory on which I stood, were the other three ships, their spars and rigging in most sorry case. The flagship, I observed, was the one anchored nearest the point of the headland, indicating that Loralie had escaped death, for which I was deeply thankful. From where I stood I could see the crews of the three ships busy repairing the damages which the storm had wrought.

Crouching in order that I might not be observed, I made my way back into the gulch, where most of my fierce retainers were finishing their morning meal.

"The food–men have returned," I said. "Keep out of sight so they will not know that we are here. And do not go far away, as I will probably need you to fight very soon."

"We will remain nearby, Rogo," said Rorg. "We are all very hungry for the flesh of food–men."

I returned to my lookout on the rock and tried to formulate some plan of attack. Presently I saw two scarlet–clad figures appear on the deck of the flagship. The smaller of the two was constantly attended by two armed warriors. Gadrimel had evidently found it expedient to keep the princess under constant surveillance.

But a plan did not suggest itself to me until I saw several boats lowered and a party of officers, headed by Gadrimel, put off for shore. Dashing back to the gulch where my cave–apes were grouped, I said, "Some of the food–men are coming ashore. We will divide into two parties of equal size, one of which will be under the leadership of Rorg. The other I will lead.

"Bores party will go down near the shore at the spot toward which they are coming. With his warriors he will climb into the trees, taking care lest the food–men see any of them, for they carry magical clubs which can kill at a great distance. As soon as the food–men enter the forest, Rorg and his warriors will drop down on them from the trees and surprise them. They can thus be slain before they have a chance to use their magic clubs. Do you understand, Rorg?"

"I understand, Rogo," replied the old cave–ape. "The food–men will not see us until we fall upon and slay them."

Calling the other cave–apes to follow me, I hurried to the other side of the promontory and descended its steep seaward side where we were hidden from view of the ships. Then, cutting the string I had with me into appropriate lengths, I tore a number of fronds from a wide–leafed variety of bush–fern, and proceeded to bind these to the heads of my subjects, spreading them in such a manner that at a distance they would effectually conceal the heads and shoulders of the great brutes. Disguising myself in the same manner, I led my savage followers to the very point of the promontory and into the water.

"You will all keep close together in the water," I said, "and follow me without noise. There are many trees and branches floating down the river this morning, and if we swim carefully and silently we will not be noticed."

Peering around the point, I saw that Gadrimel and his hunters had landed and were starting into the forest. Then there came to me faintly the yells of startled men and the roars of fighting cave–apes, interspersed with the popping of torks and clash of weapons, and I knew that all eyes on board the ship would be directed toward the scene of battle.

"Now," I said, and plunging into the water, swam around the point and straight for the flagship. Just behind me, in such close formation that we must have appeared like a single, tangled mass of floating branches, came my camouflaged apes.

The flagship was not more than a thousand feet from the point, but before we could reach it I saw more boats put off from all the ships and make swiftly for the scene of combat on shore.

We came up under the prow of the ship just as the sounds of conflict announced the arrival of the small boats at the beach where the battle was taking place.

Silently I seized the taut anchor chain and went up, hand over hand. Just as silently, my ape warriors followed. On reaching the top, I peered cautiously through the railing. Loralie and her two guards were standing on the starboard side watching the battle on shore. There were three men aloft, apparently there to repair the rigging, but they, too, had their eyes trained shoreward.

Without a sound, I climbed over the railing, and with sword in one hand and dagger in the other, advanced toward the two men. Simultaneously, I jabbed the point of my dagger in the back of one, and the point of my sword in the other.

"One false move," I said, "and you die. Raise your hands above your heads and keep your faces shoreward."

They complied with alacrity. With a little scream of fear, Loralie turned to see what had happened.

"Zinlo!" she exclaimed. "I knew you would come!"

"Take their weapons, my princess."

She quickly removed their belts from which depended their torks and scarbos.

Three of the apes had meanwhile scrambled aloft after the men in the rigging, and the others were searching the ship.

"Bring to me alive those who do not resist," I shouted. "You may slay the others."

My words had the desired effect on Gadrimel's men, for although those in the rigging all carried short scarbos, none offered to fight. Other than these three and the two I had disarmed, the apes found only the cook and his helper.

When the prisoners had all been rounded up, I addressed them.

"All of you who are willing to take orders from me will give the royal salute. The others will be quickly deposed of, as my apes are hungry."

To a man, they saluted.

"You three," I said, addressing the men who had been aloft, "hoist the sails. And you," pointing to the two guards, "heave the anchor."

I sent the cook and his helper back to their pots and pans under guard of two apes. Then I took the helm with Loralie at my side and as the sails filled, steered for the open sea.

We had nearly passed the point of the promontory when the boom of a mattork and the sing of its shell through our rigging announced that we had been discovered.

"Can you steer?" I asked Loralie.

"Better than you, landsman," she answered laughingly. "Give me the helm."

Her father ruled the greatest maritime nation on Zarovia.

"Make for the open sea," I said, "and I'll see if my marksmanship is better than my steering." The mattork, which was nothing but an oversized tork mounted on a tripod, stood nearby swathed in its water–proof covering. Beside it was the case which contained the clips of projectiles with their various designations printed in patoa: Solid, Paralyzing, Deadly, Explosive.

Stripping the cover from the weapon, I chose a clip of explosive projectiles and inserted it in the breech. By this time two mattorks on each of the anchored ships had opened fire, and shells were screaming around us. One snapped a shroud, and I ordered a sailor up to replace it. Another burst against our hull. And still others, ricocheting from the surface of the water, whined plaintively as they sped on their way.

I took careful aim at the rear mattork on the nearest ship and pressed the button. But the weapon was strange to me, and equally strange was the experience of firing a projectile from a ship. I saw my shell strike the water far behind the mark.

Again I took aim, this time allowing for the rocking of the ship. To my surprise, my shell burst just beneath my target, tearing the gunner to shreds and knocking the weapon from its tripod.

I tried another shot at the forward mattork, but it went wild. Then both boats slipped from our view as we rounded the promontory.

"My marksmanship is as wretched as my handling of a boat," I said. "But they cannot harry us for a time, at least. Where to now, my princess?"

With one hand she reached for my own, drew my arm around her slender waist. The other still skillfully managed the helm.

"Whither you will, beloved," she replied. "Shall it be Olba or Tyrhana—north or south?"

"Which is nearer?"

"They are about equally distant from here."

"Then let us try for Olba, for there I am sure Gadrimel dare not follow us."

Gently she brought the boat about until its prow pointed directly north. "It will not be long before Gadrimel sets out after us."

"He may have been slain by Rorg and his apes."

"Not he," replied Loralie. "I was watching from the ship, and saw that he was the first to run for the beach when they were attacked. Standing beside a boat and ready to put off at a sign of a turn in the tide of battle, he used his tork, but did not get into the thick of the fight. A cautious youth, my cousin."

It was not long before her prediction was fulfilled. One of the ships nosed around the promontory and came after us with all sails up.

I sprang to the mattork and fired. It was a bad miss. Again I fired. This time my projectile struck the water close to the target. I was getting the range. But when I would have fired a third time there was an explosion in the breech. The projectile had jammed and the safety plug had blown out.

Frantically I worked with the recalcitrant weapon, momentarily expecting a volley from our pursuers. But none came. Evidently the prince had forbidden the use of mattorks because of the presence of Loralie on our vessel.

Suddenly a terrific explosion from the front of our vessel knocked me flat. Half dazed, I gripped a leg of the tripod for support just as the deck gave a violent lurch forward.

My prostrate body swung halfway over, and I saw with horror that the front end of the ship had been completely blown away and she was plunging into the waves, nose down. I have never learned the cause of that explosion, but believe that the cook or his helper found a way to outwit their ape guards and destroy the vessel.

My gaze flashed to the wheel, but the princess was nowhere in sight, then I heard a shout from the water behind me. Loralie was swimming in the wake of the swiftly sinking vessel. "Jump!" she cried. "Jump quickly, or you will be dragged down with the ship!"

I sprang to the rail and leaped over. A moment later I was swimming beside her as we both strained every muscle in our endeavor to put as much distance as possible between ourselves and the stricken vessel before she went down.

But try as we would, we could not escape the mighty suction of the boat as it plunged beneath the waves. Like tossing corks we were dragged back in spite of our utmost efforts. But by the time we reached the center of the whirlpool it had so far subsided that the water was comparatively calm and we were not drawn under.

Presently bits of wreckage began to come up around us. A huge timber suddenly popped to the surface. We swam to it and found it amply buoyant to sustain our combined weight in the water.

As we topped the crest of a wave I glanced back. The first ship was within a quarter of a mile of us, and I caught a glimpse of a scarlet–clad figure in the bow, eagerly scanning the water with a glass.

I was still looking back when a cry from Loralie attracted my attention in another direction. "A killer norgal! The scourge of the Ropok has seen us! We are doomed!"

Bearing down on us at terrific speed, I saw an enormous fish. Its body, fully thirty feet in length, was blue in color, and bristled with sharp spines of a deep crimson shade. Its huge jaws, large enough to have swallowed ten men at a gulp, were open, revealing row on row of sharp, back–curved teeth.

"Better that than Gadrimel," said Loralie with a shudder, "for we can die together. One last kiss, beloved, for it is the end."

Our lips met and clung, across the timber. Then I drew my sword, puny weapon indeed with which to meet such an enemy.

Chapter XIII

As WE CLUNG to the timber there in the tossing waves, Loralie and I, the killer norgal swiftly surged closer and closer. There was no mistaking its purpose. It had seen us and singled us out for its prey.

Suddenly a dark shadow fell on us from above. A shot rang out, followed by a muffled explosion. Where the gaping mouth of the fish had been was only a bloody mass of flesh and bone. The mighty carcass lurched, flopped about for a moment, and then turned belly upward.

Above us loomed the great bulk of an aerial battleship, swiftly descending. It hovered only a short distance above our heads. A door opened in the side and a flexible metal ladder was lowered to us. I helped Loralie to mount, then went up after, hand over hand.

An officer in the uniform of Olba helped me into the ship. He was the mojak, or captain of the vessel.

Then he bowed low with right hand extended palm downward, as did every other man in sight. "Your name, officer," I said.

"Lotar," he answered, "at your highness's service."

"Lotar, you will find quarters for Her Highness Loralie of Tyrhana, then start immediately for the Imperial Palace at Olba."

"I hear and obey," he replied, and dashed off to give the necessary orders.

We mounted to the rear turret, the princess and I, and watched the two ships of Gadrimel fast disappearing from view. Why he did not fire at us I have never learned. Possibly because the princess was on board, but more probably because he feared the powerful mattorks of the mighty Olban airship.

The princess presently retired to her quarters to rest, and I went forward with Lotar, who was directing the pilot in the first turret. "How long should it take us to get to Olba?" I asked.

The young mojak consulted his charts and instruments for a moment.

"We should be able to make the palace by nightfall, Highness," he said. "This ship is rated at a rotation."

A rotation, I recalled, meant the speed at which Venus turns on her axis, approximately a thousand miles an hour.

"Who sent you after us?"

"Your Highness's father has had the entire air fleet of Olba scouring the planet for you since your disappearance from the Black Tower. His Majesty assigned a patrol zone to each ship. I have been flying above this zone for many days. Attracted by the explosion which destroyed your ship, I flew over to investigate. With the aid of my glass I saw you and Her Highness in the water, and the norgal swimming toward you. As a marksman I have won many prizes in tournaments with the mattork. It was a simple matter for me to kill the norgal with an explosive projectile."

"It was excellent shooting," I said, "and it not only saved my life, but a life that is infinitely dearer to me. You will not find me ungrateful."

"My greatest reward lies in the knowledge that I have saved your highness for Olba. There will be great rejoicing throughout the length and breadth of the empire when the people learn that you are alive. And greatest of all will be the joy of His Imperial Majesty, Torrogo Hadjez."

For some time I strolled about the ship, examining her armament and admiring the luxury of her appointments. Presently, Loralie came out of her stateroom. We went to the salon, where hot kova was served to us in jewel–encrusted golden cups.

Night fell just as we flew above the great crescent–shaped harbor of Tureno, and its myriad lights flashed on as did those of Olba. I caught a fleeting glimpse of the lighted windows of the Black Tower as we hurtled past it. Then the pilot gently slowed the ship until we were directly above the Imperial Palace.

As we dropped toward the flat roof a number of guards came running toward us. Two of them seized the ladder which we dropped and held it while the princess descended. Then I followed.

A mojak in the uniform of the palace guard stepped up and tendered the royal salute. "His majesty will be overjoyed, highness. It was his command that I bring you before him as soon as you arrive."

There was something strangely familiar about the features of this officer. I tried to place him as he conducted Loralie and me down the telekinetic elevator.

When it stopped he bowed us into a spacious hall which led to a great, arched doorway hung with curtains of scarlet and gold, at each side of which stood two guards armed with torks, scarbos and long–bladed spears.

The four guards bowed obsequiously as we came up. Then two of them parted the curtains and there stood before us another individual whose face seemed strangely familiar to me. Yet he wore the pompous uniform of a torrango, or prime minister, which I recognized from my studies, and I knew I had never met the prime minister of Olba.

As soon as he saw me, he bowed low with right hand extended palm downward. "His Majesty the Torrogo bids you welcome. Whom may I announce as accompanying you?"

"Her Highness, Loralie. Torrogina of Tyrhana," I replied.

He bowed once more and departed. A moment later I heard him announcing our names and titles. Then a voice, which also seemed familiar to me, said, "You will conduct them before the throne."

As we followed the prime minister into the large and magnificent throne room of Olba, more guards saluted and fell in behind us. A guard of honor, I thought.

I had never seen Torrogo Hadjez, and was curious for a look at his face, but restrained my impatience until Loralie plucked at my arm.

"Look!" she whispered. "Look who sits upon the throne!"

I raised my eyes, and the features of my arch–enemy, Taliboz, leered down at me. For a moment I was stunned as I saw him sitting there, arrayed in the royal scarlet and wearing the insignia of the Torrogo of Olba. Then my hand flew to my sword hilt and I sprang forward. But before I could take a second step strong arms pinioned my own from behind and my weapons were wrested from me.

"I trust," Taliboz said, bowing to Loralie, "that you will excuse this poor reception, but as your coming was unexpected we were totally unprepared to greet you with the pomp and circumstance due visiting royalty." He turned to his minister. "See that suitable apartments are prepared for Her Highness of Tyrhana at once and conduct her there, Maribo. And Vinzeth," he said, addressing the mojak who had conducted us to the throne room, "you will also conduct Torrogi Zinlo to the suite that awaits his coming."

"You fiend!" said Loralie, facing him with flashing eyes. "What are you going to do with the prince!"

"Have no fear, Your Highness," responded Taliboz. "No harm shall come to him. Not now, anyway. Later, his fate shall rest in your fair hands."

I was dragged out a side door by two guards.

They took me down a small elevator which, it seemed to me, traveled into the very bowels of the planet before it stopped. Then I was jerked out of the car and pulled along a narrow, dimly lighted passageway that seemed to have been hewn from solid rock, until we came before a door of massive metal bars.

One of the guards produced a key with which he unlocked this door, and I was flung inside with such force that I fell sprawling on a cold stone floor and the door clanged shut behind me.

Scarcely had I fallen to the stone floor of the dungeon cell into which I had been hurled, when a shadowy form darted from its dim interior and was helping me to my feet.

"Are you hurt, Highness?" the man asked solicitously. I recognized the voice instantly, though the features were still indistinguishable to me, my eyes not having become accustomed to the semidarkness.

"Lotar!" I exclaimed. "What are you doing here?"

"I was placed under arrest with all my officers and crew immediately after you left with the villainous Vinzeth. So far as I know, my men are confined in the cells around us."

"But what is the meaning of it all? Where is the Torrogo Hadjez, and how did Taliboz attain the scarlet and the imperial throne?"

"At the time of Your Highness's disappearance from the Black Tower, Taliboz and a number of his henchmen disappeared also," said Lotar. "A short time ago he returned alone, disguised as a merchant of Adonijar and driving one of the swift mechanical vehicles which are manufactured in that country. His disguise was penetrated by a soldier of the imperial guard, who placed him under arrest and took him before Torrogo Hadjez.

"His Majesty questioned Taliboz about your disappearance, and he told a story which was believed by some and discredited by others—namely, that there was a plot on foot among the guards of the Black Tower to assassinate you as you slept. He said that he, with Vinzeth and Maribo and his men, had fought, protecting you from death, until they were driven back, and you were dragged to the tower top and spirited away by the plotters in one of the tower airships.

"As quickly as he could, so his story went, he returned to his fighting craft and set out in pursuit of your abductors. They finally crashed, he said, in the wild country of the cave–apes beyond Adonijar, where you and your abductors were killed in the crash. All of his men were killed and eaten by cave–apes, and he barely escaped with his life to Adonijar, where he had purchased a merchant's outfit and vehicle with which to traverse the high road to Olba."

"I have met liars," I said, "on three planets, but Taliboz seems to be prince of them all. This, however, does not explain how the traitor attained the throne. I left him, paralyzed by a tork projectile, in a forest near the mountains of the cave–apes. That he escaped the perils of the jungle is little short of miraculous."

"No one could disprove the story told by Taliboz," Lotar pointed out, "as everyone in the Black Tower had been slain. Torrogo Hadjez could do nothing but thank him for attempting to save your life, reward him with costly presents, and restore to him all the honor and authority which had been his before his departure. That the Torrogo did not believe his story, however, was evidenced by the fact that his air navy continued to patrol the globe in search of Your Highness."

Someone rapped sharply on one of the massive bars of the cell door with the hilt of a weapon. It was one of the guards assigned to patrol the corridor.

"Less noise in there, prisoners," he growled, then passed on.

"I learned more while we were being held in one of the upper rooms after our arrest on the palace roof," continued Lotar softly. "As you are probably aware, every man who awaited us on the roof was a henchman of Taliboz. Your Imperial father, Highness, died at the hands of an assassin several days ago. The dagger found driven in his back was proved to be that of Arnifek, his prime minister. With Torrogo Hadjez dead and your highness presumably so, there was no successor to the throne and it was necessary for a new Torrogo to be elected by acclamation. Taliboz was thus elected. He immediately had Arnifek, the supposed assassin, executed, made Maribo his prime minister, and Vinzeth captain of the palace guards."

"Do you think Arnifek was guilty of the murder?"

"Of course not. Taliboz—or one of his tools—did it with Arnifek's dagger. It was part of his plan to get control of the Olban government. Why he has let you live even this long is a mystery to me."

"It is no mystery to me," I answered. "He dropped some hint of his purpose before he sent me from the throne room, for I heard him tell Princess Loralie that my fate should rest in her hands. He will attempt to force Loralie into marriage with him by threatening my life—and have me slain once the marriage is consummated."

"You are right, Highness," said Lotar. "Taliboz plays for even greater stakes—to unite the only air power and the mightiest maritime nation of Zarovia, Olba and Tyrhana, by marriage. Adonijar would probably form an alliance with him because her ruler is married to the princess's aunt. He would be the wealthiest and most influential monarch on the globe. Nor is there a single nation powerful enough to oppose such a strong alliance—not even Reabon, with her mighty army. Reabon is far across the ocean, and besides, her great warlike Torrogo died recently, leaving his daughter, Vernia, to rule in his stead."

"Reabon," I mused. "The name sounds familiar. Ah, I remember. That is the country to which Grandon went."

"Grandon?" he exclaimed, puzzled. "The name has a foreign sound."

"An old friend of mine. You would not know him. He is, as you say, a foreigner…Is this Taliboz so popular that the people would gladly make him Torrogo by acclamation?"

"Far from it, Highness," replied Lotar, "though he probably persuaded some of them to espouse his cause by convincing them that he had risked his life in an attempt to save yours."

"It looks," I said, "as if it were impossible to escape from here."

"I am familiar with these dungeons, Highness, as I served in the palace guard for two years. There is a way to escape—a secret way which I doubt very much whether Taliboz himself knows. But we must first get past yonder barred door and the armed guard in the corridor."

"If that is all," I replied, "I see freedom in the offing. Follow my instructions implicitly, and we'll soon be out of this."

"You have but to command, Highness."

"Very well. When next the guard approaches on his rounds, talk very loudly. No doubt he will stop and order you to be silent. When he does this, insult him."

"But he will only come in and beat me with the flat of his scarbo, Highness."

"Do as I say, Lotar. I will attend to the rest."

It was not long before we heard the heavy footfalls of the guard in the corridor. I immediately started a conversation with my companion in a loud voice.

"Silence!" roared the guard. "The other prisoners want to sleep."

"Be on your way, you clumsy lout," replied Lotar, "and do not in the future forget how to address your superiors."

"My superiors! Ho, ho!" jeered the guard. "Very soon will I show you who is superior, a prisoner or his jailer."

He took a bunch of keys from his belt pouch and fumbled among them until he found the one that fitted our door.

"Now see what you have done, Lotar," I exclaimed, simulating great fear. "You have got us a beating with that noisy tongue of yours."

The guard flung open the door, a grin of delight on his features. Such a man would not only welcome any opportunity to torture a fellow creature, but would seek such an opportunity.

"So, O cub of a dead marmelot, you fear a beating," snarled the guard. "It is well that a weakling such as you can never mount the throne."

"Were he on the throne," Lotar snapped, "hahoes like you would be working in the quarries where they belong!"

The guard raised his scarbo for a heavy blow at the defenseless Lotar. This gave me the opening for which I had been waiting. With a single bound I was in front of him. Before he could recover from his surprise I planted a crashing right hook on the point of his jaw. He went down like a felled ninepin, nor was a second blow necessary.

I gave his tork and dagger to Lotar, but retained the scarbo myself. It took us but a few moments to bind and gag the prostrate guard with the straps of his own accouterments. We dragged him back into a corner, closed and locked the cell door, and tiptoed stealthily down the corridor, the young captain in the lead.

"Let us release your men," I said.

"Your Highness's life is too precious to risk for them. Still, if it is your Highness's command..."

"It is."

Pausing before the first cell door, Lotar peered within.

"Here are six of them," he whispered, testing his keys in the lock.

Looking over his shoulder, I saw six shadowy forms on the floor, and could hear their breathing as they slept.

When he had found the right key, Lotar opened the door quietly and stepped within. One by one he awakened the sleeping men, cautioning silence.

We went from cell to cell until we had released forty–five men—all but five of the crew of Lotar's aerial battleship. He was opening their cell door when we heard the clatter of footsteps, the clank of weapons and the sound of talking. Armed men were approaching by way of a transverse corridor.

"Quick, into this cell, every man of you," I ordered.

Silently our forty–five filed into the cell with the remaining five. When all were inside there was standing room only.

"Now, Lotar," I whispered, "let us go to greet our callers."

He whipped out his dagger and followed me to the intersection of the two corridors, where we crouched, breathlessly awaiting the approach of the enemy.

Chapter XIV

As LOTAR and I crouched against the corridor wall in the dungeon beneath the Imperial Palace of Olba we could hear our unseen enemies drawing nearer and nearer in the transverse passage way. How many there were, or how well they were armed, we had no means of knowing. But we were desperate, and had there been an entire company of them we could have done nothing but fight like cornered rats.

Two guards, fully armed, suddenly rounded the turn facing us. Out came the scarbo of the one nearest me, but before he could use it my point had found his throat. He went down with a queer gurgling sound. Lotar had, meanwhile, sprung on the other guard like an enraged marmelot, burying his dagger in his breast. Simultaneously, we withdrew our dripping weapons, thinking this was all, when suddenly a third guard rounded the corner.

This time we had no element of surprise in our favor, for he had seen us as quickly as we had him.

He quickly clapped his hand to his tork, at the same time raising his voice to alarm the guards. "Help! Two pris—"

He said no more, nor had he even an opportunity to press the tork button, for with lightning quickness that the eye could scarce follow, Lotar had hurled his bloody dagger straight at the enemy's face. It entered his opened mouth with such force that the point protruded from the back of his neck and the hilt clicked against his teeth. With a look of amazement and horror on his twisted features, he slumped to the floor.

"Get their weapons, Lotar," I ordered, and hurried to summon our men. With the weapons of the three guards we partly armed six of them, and once more hurried away under the guidance of Lotar.

But we had not gone far when there was a great clamor and much shouting behind us, and we knew our escape had been detected. We bounded forward now, without any attempt at silence. A moment later Lotar called a halt before a huge, cylindrical pillar about three feet in diameter, which to all outward appearances was exactly like the many other pillars which supported the stone roof of the corridor.

Whipping out his dagger, he pressed the point into a tiny crack in the floor in front of it, whereupon, much to my amazement, I saw that the pillar was turning quite rapidly, and as it turned, moved up into the rock above it like a gigantic screw. In a few seconds its base was above the floor, and beneath it there yawned a black well.

"Into it, every man of you, quickly," ordered Lotar.

The man nearest the wall paused gingerly on the edge.

"Leap," ordered the captain. "It is not far."

In he went, and we could see that the spot where he had landed was scarcely seven feet below the floor level. After him, as fast as they could find room, crowded the other men. But meanwhile, the sounds from behind us told us that our pursuers were dangerously near.

It seemed an age before the last man leaped into the hole, followed quickly by Lotar and me.

Stooping down, the young mojak pressed a lever in the floor. The pillar started downward, the direction of its turning reversed, and soon we stood in total darkness. Judging from the sounds above, the thing had been accomplished just in time. The large party of guards above clattered on past without even stopping to investigate.

"They do not suspect," said Lotar, "which is well. It may be that we shall want to pass this way again. Come, I will lead the way."

As none of us had the means to make a light, we moved forward like blind men, following the voice of Lotar, who seemed to know the way by heart. "A steep slope ahead," he would sing out, or, "A sharp turn here. Look out for it." We followed him in the inky blackness.

The tunnel had apparently been hewn through the rock stratum that underlay this part of Olba. How it was ventilated I had no means of knowing, but though the air was cool and moist it seemed quite fresh.

When we had traveled for more than an hour in this fashion, I asked Lotar how much farther we had to go.

"We are but a third of the way, Highness," he responded. "This tunnel leads to the Black Tower."

"And whom do you expect to find in the Black Tower?"

"Friends. It is hardly likely that Taliboz has manned it with his henchmen so soon, but even if he has, some of us are armed and we have the advantage of surprise on our side."

"Unless," I observed, "he discovers that we have come this way and sets a trap for us."

"It is not likely. The guards in the dungeon were completely baffled. By now I doubt not that the traitorous Taliboz is exceedingly mystified and furiously angry."

It was nearly ten Earth miles from the Imperial Palace to the Black Tower, so that, traveling blindly as we were, it took us more than three and a half hours to make the trip.

When we reached our destination, Lotar cautioned silence and groped about in the darkness for some time. Then I heard the click of a lever and the turning of a cylinder, and presently a circle of light appeared above our heads—most welcome after three and a half hours of intense darkness.

Gripping the edge of the floor, Lotar drew himself up and peered cautiously about. Evidently satisfied that he was unobserved, he clambered on out of the hole, beckoning to us to follow. It was not long before we had our entire company lined up in a large room, the ceiling of which was supported by pillars similar to the one which had been raised to let us in. Lotar then pressed the hidden button that started the pillar rotating in the opposite direction, and watched it turn back into place, leaving no sign of the way by which we had come.

There were three windows in the room through which the first faint streaks of dawn were visible. There were also three doors. Lotar slowly and carefully opened one of these. But scarcely had he looked out ere a sharp challenge was hurled at him from the corridor.

"Move and you die! Who are you?"

"Lotar, Mojak in the Imperial Air Navy," replied the young officer.

"What do you here?"

"That," replied Lotar, "I will tell your mojak if you will fetch him. Who is in command here?"

"Pasuki commands," replied the guard.

"A good and loyal soldier. Take me before him."

He motioned with his hand for us to remain in the room. Then he stepped out, closing the door after him. Evidently the guard had not the slightest suspicion of our presence.

Not more than ten minutes elapsed ere the door opened once more and Lotar entered, followed by a tall, straight, white-bearded man who wore the uniform of Mojak of the Black Tower Guards, easily distinguished by the small replica of the tower worn on the helmet and the same device in relief on the breastplate.

The old soldier bowed low with right hand extended palm downward.

"Pasuki is yours to command as of old, Highness," he said, "and overjoyed that the report of Your Highness's death was false."

I did not, of course, remember Pasuki, but it was quite evident that he remembered the former Zinlo. "You were ever a true and loyal soldier, Pasuki," I replied. "See that these men I have brought with me are fed, housed and armed."

After a brief order for the disposal of Lotar's men to a mojo who waited outside, Pasuki conducted us to the telekinetic elevator and by it to my apartments.

"I'll send for you men soon," I told them. "Meanwhile we must try to devise some plan of attack on this wily Taliboz, and find a way to rescue Her Highness of Tyrhana."

Pasuki and Lotar bowed low and withdrew.

After a bath and a change of clothing, I was served with the usual huge and variegated breakfast with which Zarovian royalty tempts its appetite, to the accompaniment of gold service and scarlet napery.

But ere I had completed this meal, a page came to announce that a man who had just been admitted to the tower, craved immediate audience with me. "Who is he?" I asked.

"He gave the name of Vorvan to Pasuki, who questioned him and seemed satisfied of his loyalty," replied the page.

"Then show him in," I answered. The name Vorvan had a familiar ring, and I was trying to remember where I had heard it before when a man clad in the conventional blue garb of a tradesman entered.

He appeared about fifty years of age, and his square–cut beard had an unnatural reddish tinge, as if it had been dyed. His eyebrows were similarly treated, and a bandage was drawn across one cheek and the bridge of the nose, as if he had been recently wounded. I could not remember ever having seen the man before, yet there was something about him that was strangely familiar.

He bowed low, right hand extended palm downward.

"I have a message for Your Highness's ears alone," he said, with a significant look at the three men who were serving my breakfast.

"Won't you have some breakfast?" I asked.

"With Your Highness's leave I will decline, as I have already breakfasted. There is much to be done, and time presses." Again he glanced impatiently at the servants.

With a wave of my hand, I dismissed them.

"The page told me you gave the name of Vorvan," I said when they were gone. "Both the name and yourself seem somehow vaguely familiar, yet I cannot remember having heard it, nor having seen you before."

"Then my disguise must be effective, Highness," he answered, with a smile which was also familiar. "I am Vorn Vangal."

The smile and the name instantly brought a flood of recollections. This was indeed Vorn Vangal, the man who had arranged with Dr. Morgan to bring me to Venus–Vorn Vangal, the great nobleman, scientist and psychologist of Olba—the man who had welcomed me to Venus with the identical smile he was now wearing.

But at that time he had been attired in the purple and the glittering bejeweled panoply of a great noble, and his beard and hair had been iron gray. A bit of dye, a bandage, and the clothing of a tradesman had wrought vast change in his appearance.

"I'll try to answer Your Highness's questions in due order," Vorn Vangal said. "I returned from Reabon one week after I left you in the Black Tower, expecting to find you here, safe and sound. You may imagine my astonishment when I learned that you and Taliboz had disappeared, that your guards had been slain, and that a number of dead henchmen of Taliboz had been found here.

"I immediately established telepathic communication with Dr. Morgan who was to keep in constant rapport with you, and from him I learned what had happened to you. Then I went to Torrogo Hadjez and persuaded him to patrol the area where it might be expected that you would be found. You were moving about so much that it was impossible for the airships to find you in any specific location I might name. Part of the time you didn't know where you were, hence your subjective mind could not inform Dr. Morgan, and through him, me.

"Of course I knew the report of Taliboz was a lie when he said you had been killed, but I did not dare to so inform Torrogo Hadjez. He would have demanded to know the source of my knowledge, which would have forced me to disclose the fact that his son was on your world and you were taking his place here.

"I decided to personally conduct a search for you in an aerial battleship, and Torrogo Hadjez provided me with one for the purpose, but we encountered a terrific storm before we had gone far, and the ship was forced to land, hopelessly crippled, near the Olba–Adonijar border. I immediately took a motor vehicle back to Olba, but was placed under arrest as soon as I entered the city gates, for Torrogo Hadjez had been assassinated and Taliboz was on the throne.

"He condemned me to die as a traitor, and confiscated my city palace as well as my lands, estates and treasure. With the aid of a few faithful friends, I managed to escape before his sentence could be carried out, disguised myself as a tradesman, and came here, having learned through Dr. Morgan that this was where you were to be found."

"And now," I asked, "have you any plans for rescuing the Princess Loralie and disposing of Taliboz?"

"The only method I can think of will be a bloody revolution. Most of the men who garrison the palace and the city are men of the usurper. The men who previously filled these ranks have been sent to work on and guard the private estates of Taliboz, far to the north of Olba. If we were to proclaim your return, Taliboz would immediately denounce you as an impostor, a price would be placed on your head, and you would be hunted by every military man under his command.

"The best way, I believe, will be for you to remain here until I can arouse the patriotic citizens of Olba, secretly telling them of your presence here. You can then come to Olba in disguise, and we can make a concerted effort to capture the palace and do away with the traitor who sits on the throne."

"But that will take considerable time," I said, "and in the meantime, what of Loralie?"

This question went unanswered, for at this moment one of my guards entered with the statement that Pasuki and Lotar craved immediate audience as they had a communication of the utmost importance.

"Admit them," I said.

Both saluted hurriedly as they came in, and seemed greatly agitated. "Your Highness's presence here has been discovered," said Lotar. "We must get you away at once."

"I am sorry to inform you that there must have been a traitor among my men," said Pasuki, "planted there, no doubt, by Taliboz to spy on my doings. One of my faithful servants, however, was watching Taliboz, and has dispatched a messenger to me with the information that the usurper has mobilized an army of five thousand men who are already marching on the

Black Tower."

Chapter XV

As I SAT facing the three men, Pasuki, Lotar, and Vorn Vangal, all faithful to me, but with no plans for meeting the emergency created by the advance of the army which was ten times the strength of the garrison of the Black Tower, an idea came to me.

"Will Taliboz accompany the army, Pasuki?" I asked.

"It is probable, Highness, but I cannot be certain."

"How many men in your garrison?"

"Four hundred and fifty, not counting Lotar's fifty. We could not hold the tower long against the attack of five thousand. It is best that we disband the garrison and make our escape in the flyers on the roof of the tower. There are two there, each of which will carry two men."

"But what of the princess? If you men and your followers are willing to fight both for her and for me, I have a plan—a precarious one, but possible of execution—for saving her and dethroning Taliboz."

They pledged their loyalty.

"Very well," I said. "Prepare, then, all of you, to obey my orders without question. They may seem strange to you, but if they do, remember that they are designed to outwit Taliboz. You, Pasuki, will prepare for the defense of the Black Tower at once with all your mattorks and men. You, Lotar, will keep your men armed and ready for my call, but out of sight. See that every one of them is provided with a portable light, and that there are several extra lights. Vorn Vangal will remain at my side for the present."

The two men hurried away to carry out my commands, and I leisurely finished my breakfast, while Vorn Vangal kept anxious watch out the window.

"They draw near, Highness," he said excitedly, "and Taliboz is with them, for I see the personal standard of the Torrogo in their midst."

"Good." I went to the window. Taliboz was bringing up a mighty host indeed, compared to our small garrison. When they were within a thousand yards of the walls that surrounded the tower, they deployed to the right and left. A man bearing a banner on which was written in large letters the Zarovian word "dua"—which, under the circumstances meant, "a truce"—left the ranks and marched toward the main gate of the tower wall.

"A herald," said Vorn Vangal. "Taliboz would treat with us."

"Let us go to the top of the tower."

We quickly took a telekinetic elevator.

"We are completely surrounded now," said Vorn Vangal. "There will be no escape. Even if we were to try to get away in the airships we should immediately be shot down by their mattork crews."

"We are not yet ready to attempt an escape."

The herald stopped near the gates and shouted a command to Pasuki to deliver to His Imperial Majesty, Taliboz of Olba, "the usurper who calls himself Zinlo of Olba." He offered a free pardon to Pasuki and his men.

"You will return to His Majesty," replied Pasuki, "our regrets that we are unable to comply with his order, as we have no usurper in the Black Tower."

"Who is that man in scarlet I see standing on the roof of the topmost segment?" demanded the herald. "If that be not Zinlo of Olba..." He checked himself, then continued, "If that be not the usurper who calls himself Zinlo of Olba, who is he?"

"He is Zinlo of Olba. Tell that to your traitorous master, and bid him come and bend the knee to the man whose throne he has stolen." Turning contemptuously, Pasuki walked away from the parapet.

"Pasuki has played his part well," I informed Vorn Vangal. "Now, remove your disguise; if possible get rid of that villainous–looking hair dye; array your self in the purple that suits your true station, and then report to me in my apartments."

"I will carry out Your Highness's commands at once," replied Vorn Vangal, and hurried to the elevator.

I watched the herald as he picked his way through the encircling army to a point some distance behind it where a man stood, garbed in the royal scarlet, surrounded by officers and courtiers. I knew that he must be Taliboz.

Scarcely had the herald bowed before him ere he sent a number of officers scurrying toward the front lines. A mattork spoke. The shell went screaming past the tower only a few feet from my head. A second shell exploded near me, tearing away part of the battlement.

As our mattorks replied, a general bombardment started, and the soldiers of the encircling army took advantage of natural cover when it was to be had, or threw themselves flat and dug in. I judged that they planned to bombard the tower before attempting to storm it.

Shells were rattling like hail against the upper battlements when I took the elevator and descended to my apartments. Here I found Vorn Vangal, once more the great Olban noble I had first seen.

Together we entered the elevator once more and descended to the fifth underground level, where Lotar's men were mobilized. The young mojak saluted and then stood awaiting my orders. Even at this depth the thunderous sounds of the battle came faintly from above, and I could see that both men and commander longed, even as did I, to be in the thick of it. But I had other work for all, which might prove as exciting and far more dangerous.

"Have you the lights, Lotar?" I asked.

"Every man has been provided with a light, and there are several to spare, Highness."

"Then give one each to Vorn Vangal and me, and we will start for the palace at once, the way we came. Hurry!"

Lotar quickly handed us a light each, and then led us to the pillar from beneath which we had entered the Black Tower. I led the way into the pit beneath it as soon as it was raised, closely followed by Vorn Vangal, and leaving Lotar to close the entrance and bring up the rear.

Traveling with lights, it was easy to maintain a pace much faster than our previous one when we had walked in total darkness.

"How many guards do you think there will be in the palace?" I asked Vorn Vangal as he jogged along beside me.

"Normally there are a thousand constantly on duty in the palace and grounds. However, it may be that Taliboz has taken some of these with him in order to fill the ranks of his hastily organized army. If this is the case, he may have left two or three hundred, perhaps five hundred men."

"Whether there be two hundred or a thousand, we must take the palace," I said. "In either case we will be tremendously outnumbered, but we have the advantage of surprise in our favor."

When we reached the palace, I called a halt to give the men a rest, and passed back word for Lotar to come up.

As soon as he joined us, I told him my plans for taking the palace. Then I pulled the lever which operated the pillar above us, and we all snapped off our lights.

When the pillar was high enough I drew myself up and peered over the edge of the floor through the dim light of the dungeon. Only one guard was in sight, and he was walking away from me. Silently I threw a knee over the edge, stood erect, and signed for the others to follow me. When every man was out, Lotar pressed the hidden button which closed the wall.

At the suggestion of Vorn Vangal, our torks were loaded with the projectiles which paralyze for several hours but do not kill unless they happen to strike a vital spot. By using these bullets we could render our opponents helpless without actually killing them, and would not be bothered with guarding prisoners.

As Vorn Vangal had surmised, Taliboz had taken a number of the palace guards with him when he started for the Black Tower. We found only one man patrolling the corridors of the level we were on, and he was quickly put out of the way. On the next level we found two guards, and on each of the three dungeon levels above it, two. Although they were not taken completely by surprise, having heard our shots, they were easily overcome.

On the ground level, Lotar took twenty men and started out in one direction while his lieutenant took another twenty and went in the opposite direction. With the ten remaining men, Vorn Vangal and I took an elevator to the roof.

Here we found only a dozen men on guard, and quickly shot down all but one, who surrendered in terror, for he did not know that we were not using the deadly bullets in our torks. There were six aerial battleships on the roof but crews in none of them. I also noticed several small, one–man airships. One of these suddenly rose and started for the Black Tower, but Vorn Vangal leaped to a mattork and shot it down. It crashed in one of the busiest streets of Olba, drawing a great crowd and halting traffic.

Quickly searching the other airships, we found them untenanted.

By questioning the man we had captured, we found that Vinzeth, Mojak of the Palace Guards, had ordered most of his men to the dungeon, and had gone there himself to direct the fighting.

"Now, Vorn Vangal," I said when we were in control of the roof, "do you think that by spreading the knowledge of my return in Olba you can get us a few more fighting men?"

"I can raise a vast army, and that quickly. They may not all be trained soldiers, but every male Olban knows how to use a tork and scarbo."

"Then you will remain here in charge of the roof, retaining five men at all times to defend the stairway. The other five you may use as messengers to summon your friends. As all these men are from an aerial battleship, I assume that they know how to handle the small airships."

"They do," replied Vangal.

I then sent for the prisoner. When he was brought before me I asked him where the Princess of Tyrhana was imprisoned.

"I do not know, Your Highness," he replied.

"Have a care how you lie to me," I warned him.

"I swear it, Highness. I have no idea of her whereabouts."

"Cling to your falsehood, knave! We shall see if it will sustain you in mid-air. Pitch him over the battlements, men."

The two warriors who had brought him immediately began dragging him toward the battlements. He struggled unsuccessfully to break away from them, feet threshing, eyes rolling in terror.

"Wait!" he shrieked. "I know! I will tell!"

"Bring him back," I ordered. "He shall have another chance."

Once more they brought him before me, this time trembling with terror and thoroughly cowed.

"Speak," I said. "And tell the truth this time."

"Her Highness has apartments on the floor just beneath us," he said quaveringly. "The last floor at which the elevators stop."

"And how is she guarded?"

"Two men guard her door, and she has two female attendants."

I did not wait to hear more but dashed down the stairway. After traversing several corridors, I saw two guards standing before a door draped with scarlet, and knew I had the right place. One of the guards saw me as soon as I saw him, and our torks spoke in unison. His bullet struck my sword hilt, but mine stretched him, unconscious, on the floor. The other guard wheeled just in time to receive my second bullet and share the fate of his companion.

Rushing up to the doorway, I ripped aside the scarlet drape and tried to open the door, but it was locked. I quickly searched both fallen guards but could find no keys in the belt pouches of either.

Arising, I rapped loudly and called the name of Loralie.

A woman's voice answered me from within. It was the voice of my princess. "Who is there?"

"It is I, Zinlo," I replied. "Open the door, quickly."

"Zinlo, beloved!" she answered. "I had almost lost hope of your coming. But I cannot open the door. It was locked from the outside, and we have no keys in here."

"Then I'll break it down," I answered. "Stand away from it."

Backing across the corridor, I ran at the door, hurling my body against it, but it was sturdily fashioned from thick planks of tough serah wood, and my sole reward for my onslaught against it was a bruised shoulder.

Again and again I hurled myself against it with the same result.

Then I whipped out my scarbo, resolved to hew my way through it, when I suddenly heard the sound of men running behind me. Wheeling, I beheld the brutal, leering features of Vinzeth. Behind him came a dozen palace guardsmen. I reached for my tork, but before my hand touched it, his spoke. There was a soaring pain in my already bruised shoulder, a dizzy nausea swept over me, and all went black before my eyes.

When I regained consciousness after being shot down by Vinzeth, I had a furious headache, a terrific pain in my shoulder, and a tremendous thirst. I was lying on a mattress on the roof, with Vorn Vangal bending over me, holding a phial of some pungent liquid beneath my nostrils. Lotar was standing near by.

"Where is Loralie?" asked. "Have you rescued her?"

"Here, drink this," said Vorn Vangal, removing the phial from beneath my nostrils and holding a steaming bowl to my lips. "Then I will tell you." I recognized the fragrant aromatic smell of kova, and drank deeply. The hot, stimulating beverage sent the blood coursing warmly through my veins.

When I had drunk, Vorn Vangal said, "Lotar and his men not only conquered the guards stationed on every floor they came to, but defeated the fifty guards which Vinzeth took down from the roof to oppose them, driving them upward from floor to floor until only a dozen remained with their mojak. Evidently intending to get the princess and escape in one of the airships, Vinzeth retreated with his twelve men while Lotar was conquering the guards posted on the floor that is second from the top. This took only a short time, but when Lotar reached the top floor he saw Vinzeth standing over you with a scarbo, ready to give you the death blow.

"He instantly opened fire, whereupon Vinzeth transferred his attention from you to the only avenue of escape left to him—the door to the apartments of the princess. With a key from his belt pouch he succeeded in opening it and getting inside with two of his men. The others were shot down by Lotar and his warriors.

"Finding you were not dead, but only temporarily paralyzed, Lotar had you brought up to the roof by two of his men, and with the others who were with him, demanded that Vinzeth surrender and give up the princess. But Vinzeth refused to surrender, and swore that if the door were broken down the princess should be instantly slain."

"How long ago was this?"

"It occurred about three hours ago. The effect of the narcotic in the tork bullets lasts about that long."

"And she is still in there with him?" I asked, sitting up.

"What could we do, Highness? We have surrounded the room, but if we break in she will undoubtedly be slain. Vinzeth is a desperate character."

"You are right. We must find some way to outwit this Vinzeth."

"We have not been unsuccessful in other ways," said Vorn Vangal. "Already I have raised a citizen army of twenty thousand men, and more volunteers pour into our ranks constantly. The city is in the hands of the loyal commanders I have appointed, and a thousand men who are trustworthy guard the palace from roof to dungeons."

"What about Pasuki in the Black Tower? I had intended to have you send him reenforcements by way of the tunnel as soon as you could get them, but forgot it."

"In this I acted without Your Highness's command, guessing your intentions," said Vorn Vangal. "Five thousand men have already traveled to the relief of Pasuki through the tunnel. When all get there, his men will outnumber those of Taliboz. And they will have a decided advantage any time he decides to storm the tower. The twenty thousand citizen troops are mobilized near the south gate, awaiting your orders."

Just as he finished speaking a small, one-man flyer alighted on the roof. The man who stepped out looked around him for a moment, then espying our group, ran toward us.

"I have just come from Tureno," he announced. "A mighty battle fleet is in the harbor—the fleet of Tyrhana. And in the flagship rides Ad, Torrogo of Tyrhana, who demands that his daughter be delivered to him safe and sound, or he will immediately reduce Tureno and march on Olba. With him, also, are two ships, in one of which is Prince Gadrimel of Adonijar. He threatens an immediate declaration of war by his nation if his cousin, the Princess of Tyrhana, be not immediately returned unharmed to her imperial sire."

"Never mind Prince Gadrimel," I told the messenger, "but fly at once to the flagship of Torrogo Ad. Tell him that his daughter has been kidnapped by one of the mojaks of Taliboz, and we are trying to rescue her. Tell him further that if he cares to, he is welcome to land his army in Tureno, and that such citizens of Tureno as are available and can bear arms will march with him and assist him if he is bent on attacking the army of the man who abducted his daughter and usurped the throne of Olba."

The messenger made obeisance and departed.

I turned to Vorn Vangal. "Send another messenger at once to the King of Tureno. Tell him it is my command that he permit the soldiers of Tyrhana to land, and that he send as many men with them as he can gather to fight Taliboz. You will then go yourself and take command of the citizen army that waits at the south gate of the city, starting immediately for the Black Tower and surrounding the army of Taliboz, if possible."

Vorn Vangal hurried away to carry out my orders, and I swung on Lotar. "By looking over the battlements, can you point out the windows of the room in which Her Highness is confined?"

"Yes, Highness."

"First send for a long, strong rope," I commanded. "Then show me the windows—and be sure you make no mistake."

He sent a man scurrying for a rope and then went to the parapet and leaned over. I leaned over with him and he pointed downward.

"That window," he said, indicating one almost directly beneath us, "opens on the reception room of her apartment. The one to the left opens on her bedroom, the right on her bath."

At the sound of footsteps behind us we turned. Two soldiers bearing a large coil of stout rope saluted.

"Put down the rope," I ordered. "Now you, Lotar, go down in front of the door of the princess's apartment. Make a great noise, demand the release of the prisoner, and engage Vinzeth in an argument if you can. Don't do anything until you hear a commotion inside, or until I call you. Then break down the door."

With a quiet smile, which showed his full comprehension of my plan, Lotar hurried down the stairway.

Making a tight loop in the end of the rope, I brought it over the parapet at a position directly above the window which opened on Loralie's bedroom. Then, telling the two soldiers to let me down until I held out one hand for them to stop, I swung over the battlement, and with one foot in the loop and both hands gripping the rope, was swiftly and silently lowered. As soon as I was opposite the window, I signaled the men to cease lowering me. Because of the projection of the battlements, I hung about three feet from the window ledge. Below me was a sheer drop of about a hundred feet to the balcony roof of the next segment.

Gripping the rope with both hands, I worked it as a child works a swing until it began to move back and forth, first toward, then away from the window ledge. Nearer and nearer it swung until I was finally able to hook a foot over the ledge and draw myself inside. Cautiously dropping to the floor, I found the room deserted and the door closed. From beyond the door came men's voices raised in altercation.

Scarbo in hand, I tiptoed to the door and gently opened it a little way. Standing near the large central window, but looking toward the entrance to the corridor, were Loralie and her two handmaidens. Just in front of them, and also facing the door, were Vinzeth and his two men.

I had no idea whether the two maids with Loralie were friendly to my cause or to that of Taliboz, but I took a chance, and, reaching out, touched the arm of the one nearest me, then held my finger to my lips for silence. She started and gave a little cry of fear which caused me to snatch at my tork, but it went unnoticed by the three men because of the clamor in the corridor.

Motioning the girl into the bedroom, I touched her companion in a like manner, and also succeeded in getting her out of the way without noise. I then touched Loralie lightly on the shoulder. She swung on me, a furious look in her eyes, but it was instantly replaced by one of infinite tenderness when she recognized me. She went with me quickly enough into the bedroom, but when I started out again she threw her arms around my neck to detain me.

"Don't go, please," she whispered. "They will kill you. Close the door and stay in here."

I smiled, kissed her, and pushed her away.

"Lock the door after me," I said in a whisper. "In case I lose the fight, Lotar will break in from the corridor before Vinzeth can harm you."

Then I stepped out and softly closed the door after me. At this instant one of the men, turned, facing me. For a moment he stared incredulously; then he reached for his tork. But mine was already leveled at him, and I fired.

At the sound of the shot, Vinzeth and the other ruffian swung about. I shot the latter, but the mojak of Taliboz was too quick for me. Without pausing to draw a weapon, he sprang in so close that I was unable to use mine, and we went down in a heap, kicking, clawing, striking and gouging each other like a pair of wild animals.

The corridor door, meanwhile, was splintering from thunderous blows on its exterior. Although the thick serali planking was exceedingly tough, it was evident that it could not much longer withstand the terrific assault. Lotar had evidently found something that made an efficient battering ram.

All this came to me subconsciously as I fought, for I was too busy with my powerful and wily antagonist to think of anything else. Back and forth we struggled, rolling over and over, crashing against furniture and pulling down hangings, each man kept so occupied by the other that he was unable to use a weapon.

Presently I managed to get a short arm jolt to Vinzeth's jaw, which partly dazed him, and was about to repeat the process when he suddenly caught me in the solar plexus with his knee. With the wind completely knocked out of me, I sank, gasping, to the floor.

He uttered a yell of triumph, and whipping out his scarbo, swung it aloft with the evident intention of splitting my skull.

But ere he could bring it down, there was a final, rending crash from the corridor doorway, followed by the cracking of a tork. With a look of horrified unbelief on his features, Vinzeth dropped his scarbo and pitched forward on his face, his body lying across me.

Lotar quickly dragged him off me, and flung him into the corner as if he had been a sack of grain. I sat up but was unable to talk.

When I regained my speech I called to Loralie, telling her that it was now safe to open the door. Recognizing my voice, she came out and knelt beside me, pulling my head down on her breast and asking me where I was wounded.

But I reassured her, and a moment later, having managed to regain my breath, I stood up. "Man one of the aerial battleships at once, Lotar," I said. "We're going to pay our respects to Taliboz."

While we waited for Lotar to get the ship ready for flight, Loralie and I stood on the palace roof, looking toward the Black Tower.

Lotar sent us each a glass, and with the aid of these, we could watch what was transpiring.

The citizens' army which had started out from Olba was now less than two miles from the tower and spread out in an immense crescent. Marching from Tureno, and almost as close to the besieged tower, was an army almost as large as that of Olba, deployed in the same manner. On account of his low position and the rolling formation of the ground, Taliboz had not yet seen his approaching enemies. His men, who had evidently been previously repulsed, judging from the bodies that lay before the wall, were forming for a new assault on the Black Tower.

We were watching the horns of the two crescents draw together when Lotar called to me, "The ship is ready, Highness."

Chapter XVI

LORALIE AND I boarded the aerial battleship. It was the same one that had rescued us from the killer norgal and brought us to Olba, manned in part by the same crew, and commanded by Lotar.

By my command he piloted the ship to a point directly above the Black Tower, and hovered there. The armies from the north and south had, by this time, completed their encircling movement and were rapidly closing in on the unsuspecting army of Taliboz.

Zinlo of Olba, to Taliboz: You are surrounded by an army of forty thousand warriors. As the Black Tower is garrisoned with five thousand men, you cannot hope to take it. You have your choice of unconditional surrender or annihilation. If you surrender, lay down your arms and raise the "dua" pennon. If not, you alone are responsible for what will follow.

ZINLO

Rolling it up and weighting it with an empty tork clip, I hurled it down at the spot where the Imperial Standard of Olba fluttered in the wind.

With the aid of my glass I watched its flight downward, and saw it fall near one of the officers, who carried it to his commander.

Unrolling it, Taliboz read it, then passed it to the man nearest him. Upon careful scrutiny with the glass, I saw the man was Maribo, his prime minister. After the latter had read it, the two engaged in a lengthy argument in which several of the others joined.

I judged from their attitudes that the other officers sided with Maribo, and that Taliboz stood alone in whatever decision he had made. While the argument was going on, the first skirmish line of the encircling army opened fire.

Suddenly wheeling and walking away from Maribo and the others, I saw Taliboz shout something to a mattork crew and point toward our ship. A moment later a shell screamed past me. This was his answer.

A gunner in our forward turret promptly replied, wiping out the crew of the mattork from which the shot had been fired.

But Maribo and the other officers apparently did not approve of the way Taliboz had replied to our missive. With positive defeat staring them in the face, they appeared to be united in favor of immediate surrender. At least they did not interfere with Maribo when he ran up behind Taliboz just as the traitor was ordering another gun crew to fire on us, and deliberately stabbed him in the back.

Scarcely had the stricken traitor sunk to the ground ere Maribo gave an order to the standard bearer. Instantly the banner of Taliboz was lowered and the pennon of peace raised, while the shout of "dua" went around the lines. The fighting ceased almost instantly, and with their weapons on the ground and their hands clasped behind their heads in token of submission, the warriors who had set out so confidently that morning to reduce the Black Tower, were taken prisoners.

"Now that they have surrendered," said Loralie anxiously, "can't we go and see my poor father?"

"We'll get him and take him to the palace at once. I want him to be my guest as long as he cares to stay."

"And I want you to ask him something just as soon as you get a chance," she said with a meaning smile. "Remember Cousin Gadrimel is with him. He is very fond of my cousin."

We flew southward to where the standards of the Torrogo of Tyrhana, the Torrogi of Adonijar, and my Rogo of Tureno fluttered in the breeze, then descended.

As Loralie and I got down from the ship, three men came to meet us. All wore the scarlet of royalty. The foremost I recognized instantly by his mincing gait as Prince Gadrimel. The other two I did not know except by their insignia.

Loralie flung herself into the arms of the taller of the two, a straight, athletic–appearing monarch with snapping brown eyes and a square cut, jet black beard. I judged him to be about forty years of age.

"Father!" she cried joyously.

He kissed her hungrily, then held her away from him, looking her over from head to foot. "My little girl. I can scarcely believe it is you, alive and well. Rather had I lost my empire and my life than that harm had come to you."

"This is Zinlo of Olba, Father," she said, indicating me. "Prince Zinlo, my father, Torrogo Ad of Tyrhana."

"You have placed me deeply in your debt by bringing my daughter to me unharmed," said Ad.

"Had there been a debt, Your Majesty," I replied, "it would have been canceled long ago by the pleasure of Her Highness's company."

Gadrimel came up and bowed formally, muttering something about being grateful to me for having rescued his dear cousin and fiancee. The other in scarlet was the Rogo of Tureno.

I asked that he arrange for the entertainment of all soldiers and sailors of Adonijar and Tyrhana, in his city, at the expense of the Imperial Government of Olba.

Ad and Gadrimel then got aboard with us. We flew to the Black Tower, where we took Pasuki on board, and to the headquarters of the citizens' army, where we picked up Vorn Vangal. Then we flew to the palace.

When quarters had been assigned to our guests, Vorn Vangal enthusiastically undertook the task of supervising preparations for a great feast to be held that evening. I met my guests in the imperial reception room, where I ordered kova served.

Gadrimel was so attentive to Loralie that I scarcely had an opportunity to speak to her. So I called her father out on the balcony, told him I loved Princess Loralie, and asked him for her hand in marriage.

Ad looked astonished. "Beard of my grandfather!" he thundered. "What's this you say? Her hand in marriage? Is it possible that you are not aware that she is to marry her cousin Gadrimel?"

"I knew that she was betrothed to Gadrimel against her will," I replied, "but that does not stop us from loving each other."

"From loving each other! Loralie—come here, child." He added, "Excuse us a moment, Gadrimel."

Loralie came out through the window, visibly a little frightened at his tone.

"I hope," he said gravely, when she stood before him, "that you will deny, once and for all, that you love His Highness of Olba. You know my wishes with regard to Gadrimel!"

For a moment she hung her head, but for a moment only. Then she raised it proudly, and with tears brimming in her glorious eyes answered, "Father, I love him, and have told him so."

On the dark brows of Ad a storm of anger was gathering.

"By the blood and bones of Thorth!" he roared. "Do you thus defy me—me, your father? You ingrate! I swear by my head and beard that I'll wed you to Gadrimel at once and take you to Adonijar."

"Father, please!" Great tears were streaming down her cheeks now.

"Your Highness," Ad said to me shortly, "you will confer a favor on me by leaving us."

I bowed and departed, striving to conceal my bitter disappointment as I entered the room where we had left Gadrimel. The prince had a most unwonted grin on his effeminate face, and I had no doubt but that he had been listening a moment before at the window.

He instantly began a lisping chatter about our many adventures together, and his own heroic exploits after we had parted company in the fern-forest.

At intervals when he stopped talking long enough to sip his kova I could hear the sobbing of Loralie on the balcony and the rumbling voice of her father. Suddenly Ad appeared to lose his temper again, for he roared, "He did, did he? Why, of all the…"

He strode to the window, his face a thundercloud of wrath. Loralie hurried after him. I leaped to my feet, expecting physical violence.

But he did not even look at me. Instead, he walked to where Gadrimel was sitting and, seizing him by the scruff of the neck, jerked him erect.

"You insolent cub!" he roared, shaking the prince until his teeth rattled and his eyes nearly popped from his head. "You mincing, lisping, addle-headed popinjay! So you would abduct my daughter and force her to marry you! Lucky it is for you that I am constrained to remember you are the son of my sister. Were it not for that I should wring your neck and hurl you from the battlements."

"I—ah, ah, you're choking me," gasped the prince.

"Did you think I was fondling you, you wretch?" thundered the Emperor of Tyrhana, and shot the princeling through the window by applying his toe to the youth's center of gravity. Nor did he return, but slunk away through another room.

A look of serenity gradually settled over Ad's clouded brow. "Your Highness, like all men, I sometimes change my mind."

"It is a mark of greatness," I replied, bowing.

"Tonight at dinner, my children, I will announce your betrothal."

Before either of us could reply a guard entered and announced that Vorn Vangal, Pasuki and Lotar craved immediate audience.

"If Your Highness can spare a moment to the people," said Vangal, "please be so good as to show yourself on the balcony."

"What is up?" I asked.

"A little technicality to be cleared up," he answered. "Taliboz was only wounded and not killed as we thought. He has escaped. Under the law he is still Torrogo of Olba because he has been legally so acclaimed, thus taking precedence over your otherwise perfectly legal succession to the throne. Knowing all the circumstances the people of Olba now wish to acclaim you Emperor, so there will be no complications hereafter."

I walked to the balcony. The palace grounds were thronged with a close-packed, surging populace. The streets were jammed with people, and every window ledge, balcony, housetop and wall in sight was packed.

As soon as I appeared above the battlements a hundred thousand scarbos flashed aloft in the hands of the men, and a hundred thousand white scarves were waved above the mighty sea of humanity by the women and girls. A great cheer rose, swelling in volume until it seemed that it must shake the very palace.

"Hail Zinlo, Torrogo of Olba!"

I bowed in acknowledgment of this tremendous ovation, whereupon every voice was suddenly stilled.

"I thank you, my people," I shouted down to them. "I will ever strive faithfully to fulfill the trust you have placed in me."

Once more the scarves and scarbos flashed aloft. Once more a thunderous cheer rolled up. Bowing, I returned to the room and the congratulations of my friends.

With the deepest satisfaction I appointed Vorn Vangal prime minister, and gave the command of army and aerial forces to Pasuki and Lotar. My three loyal friends made obeisance and departed, leaving Loralie, Ad and myself alone.

"Since you have made so free with your favors, Your Majesty," smiled Loralie, "what have you left for me? Am I not also to be honored?"

"Why, yes," I answered, as, unmindful of her father's presence, her arms went around my neck. "As soon as you grant me leave, I'll make you Torroga, Empress of Olba."

"It's the highest honor an empire can bestow," laughed Ad, "for be he in palace or hovel a man is ever subject to the sweet will of his wife."

"Agreed," I replied. "And now, little wife to be, what is your pleasure?"

"If you were not so busy talking nonsense to Father," she pouted, "you would see that I have been waiting for you to kiss me."

Afterword

THUS ENDS the tale of Rorgen Takkor's adventures on Venus, up to the time that he was securely established as Zinlo, Torrogo of Olba. However, lest the perceptive reader remind me that this security was precarious at the very least—since Rorgen Takkor had merely exchanged personalities with Zinlo of Venus, who was meanwhile on Earth in the body of the man known as Harry Thorne—let me assure him that I have not forgotten this fact.

Robert Grandon was in exactly the same position, in Reabon, at the close of his story, which is told in "The Planet of Peril." Those who have read that story know that the resolution of Grandon's difficulty in this regard also solved Rorgen Takkor's problem. So I will only mention here that neither Grandon nor Takkor had to worry about being taken from their wives and thrones and returned to their Earth bodies; but how this came about you will have to read the novel mentioned above to discover.

The Author.

www.ingramcontent.com/pod-product-compliance
Lightning Source LLC
Chambersburg PA
CBHW071501220526
45472CB00003B/875